Civil Engineering Core Concepts Series

Understanding the Built Environment: Principles, Systems & Real-World Infrastructure Design

Martin Foster

TME Press email: support@tme.net

Introduction

Step outside this morning and count the pieces of civil engineering that got you to wherever you're reading this. The road under your feet. The water pipe running to the nearest tap. The bridge over the stream at the edge of town. The storm drain at the curb. The steel frame of the building above you. The traffic light that stopped the cars so you could cross. Every one of these things was designed by someone, built by someone, and is being maintained by someone right now. Collectively, that work is civil engineering, and it's the oldest and broadest branch of the engineering profession.

Civil engineering touches more lives, more hours of the day, than perhaps any other technical discipline. The weight of civil engineering gives the profession its seriousness and its code of conduct. It also gives it a richness that few outsiders appreciate until they start studying it. A civil engineer working on a highway interchange is juggling traffic flow, soil behavior, stormwater runoff, pavement performance, construction logistics, budget, public safety, and political pressure, often all in the same afternoon. The discipline rewards people who can think across scales, from the molecular behavior of cement paste to the regional hydrology of a watershed, and who can translate that thinking into drawings a crew can actually build.

This book is a guided tour of that discipline. It's designed for students entering civil engineering programs, professionals in adjacent fields who need a structured overview, architects and planners who want to speak the language more fluently, and curious readers who've always wondered how the physical world around them actually works. No prior engineering background is assumed. Calculus and physics help but aren't required. Where math appears, it's explained in words first, then in symbols, and only where it sharpens understanding rather than obscures it.

The structure follows the way civil engineering is typically organized in practice and in university curricula. The early chapters build the foundation.

Chapter 1 covers the profession itself, its history, its branches, its ethics, its measurement conventions, and its project lifecycle. Chapter 2 introduces the mechanics and statics that underlie every structural calculation an engineer will ever do. Chapter 3 surveys the materials of construction, from concrete and steel to mass timber and fiber-reinforced polymers. Chapter 4 brings mechanics and materials together in structural analysis and design, including how loads are classified and how safety margins are chosen.

From there the book moves outward into the major specializations. Chapter 5 goes into geotechnical engineering, the science of the ground itself, which determines what can be built on a given site. Chapter 6 covers transportation and highway systems, the networks that move people and goods. Chapter 7 addresses water resources and hydraulic engineering, from watersheds to pipe networks to flood control. Chapter 8 handles environmental and sanitary engineering, the processes that provide clean water and manage waste. Chapter 9 steps back to look at construction management and project delivery, the business and logistics of getting projects built. Chapter 10 explains surveying and geospatial engineering, the measurement backbone of every project. Chapter 11 zooms out to urban planning and integrated infrastructure systems. Chapter 12 looks at the present and future of sustainability, resilience, life cycle assessment, digital twins, and the emerging technologies reshaping how infrastructure gets conceived and delivered.

Each chapter stands on its own, so readers who want to skip around can. The chapters build on each other, though, and reading them in order will reward you with a cumulative understanding that isolated chapters can't give.

A few conventions are worth flagging up front. The book uses SI (metric) units as the primary measurement system, with U.S. customary units in parentheses where American practice uses them, which in civil engineering is often.

A word on breadth versus depth. A single book can't make anyone a practicing civil engineer. The licensure pathway in most jurisdictions takes four to six years of university, followed by a supervised practice period of typically four years, followed by professional examinations. What a book like this can do is give you the shape of the discipline. It can show you how the pieces fit together, what the major questions are in each specialization, what vocabulary practitioners use, and what distinguishes a well-designed system from a poorly designed one.

After reading it, you should be able to follow a conversation between engineers, read a project drawing with comprehension, and recognize the trade-offs embedded in any infrastructure decision. That's a valuable place to stand, whether you go on to formal study or simply want to be a more informed citizen of the built environment.

The book also tries to communicate something harder to articulate, which is the sensibility of the profession. Civil engineers are generally conservative in the technical sense. They use materials and methods whose behavior has been studied for decades or centuries. They build in redundancy. They plan for the failure mode as carefully as the normal mode. This conservatism isn't timidity; it's earned humility in the face of systems that will be used by strangers for generations. A well-designed bridge outlives the engineer who designed it, often by a lot. A water treatment plant serves children not yet born. The obligation is to them as much as to the client who signed the contract.

At the same time, the profession is changing faster than it has in a hundred years. Climate change is forcing a rethink of what loads infrastructure must withstand. Digital tools are collapsing the boundary between design and construction. New materials and manufacturing methods are opening possibilities that were science fiction a generation ago. Cities are getting denser, more instrumented, and more complex. The late chapters try to give you a sense of where these currents are taking the field, while grounding those changes in the same physical and ethical principles that have always governed civil engineering practice.

One final note on how to read this book. Civil engineering rewards intuition, and intuition is built by exposure, not memorization. As you move through the chapters, try to connect what you're reading to things you've seen. Notice the curve of the highway on your next drive and think about the geometry behind it. Look at the expansion joints on a bridge and consider what movement they accommodate. Watch a construction site and try to identify which trade is doing what, in what order, and why. The world becomes visibly more interesting when you have the vocabulary to describe it, and civil engineering, more than most disciplines, is written into the landscape for anyone willing to look.

With that, we'll cover our outline and begin where the profession begins: with its own history, and the people, problems, and principles that shaped it into what it is today.

Outline

Chapter 1: Foundations of Civil Engineering

- History and evolution of the civil engineering profession
- Branches and specializations within the discipline
- Role of the civil engineer in society
- Professional ethics, licensure, and codes of practice
- Units, measurement systems, and engineering notation
- The design-build-operate lifecycle of infrastructure

Chapter 2: Engineering Mechanics and Statics

- Force vectors, equilibrium, and free-body diagrams
- Moments, couples, and reactions at supports
- Centroids, moments of inertia, and distributed loads

Chapter 3: Properties and Behavior of Construction Materials

- Concrete composition, curing, and strength development
- Structural steel grades and metallurgical properties
- Timber, engineered wood, and mass timber products
- Masonry, mortar, and bonding techniques
- Composite materials and fiber-reinforced polymers

Chapter 4: Structural Analysis and Design

- Load classification: dead, live, wind, seismic, and thermal
- Beam theory, shear, and bending moment diagrams
- Trusses, frames, and indeterminate structures
- Limit state design and safety factors

Chapter 5: Geotechnical Engineering and Soil Mechanics

- Soil classification and index properties
- Effective stress, pore pressure, and consolidation
- Shear strength and slope stability analysis
- Shallow and deep foundation systems
- Earth retention structures and retaining walls
- Ground improvement and soil stabilization techniques

Chapter 6: Transportation Engineering and Highway Systems

- Pavement types, subgrade design, and flexible versus rigid systems
- Horizontal and vertical roadway alignment geometry
- Traffic flow theory, capacity, and level of service

Chapter 7: Water Resources and Hydraulic Engineering

Contents

Chapter 1: Foundations of Civil Engineering

History and Evolution of the Civil Engineering Profession

Civil engineering is older than the word itself. Long before anyone called it a profession, people were moving water, building walls, spanning rivers, and leveling ground, and the people who did this work well enough to be remembered left some of the most durable artifacts of the human record.

The pyramids at Giza, the ziggurats of Mesopotamia, the step wells of India, and the aqueducts of the Roman Empire all represent civil engineering in its pre-modern form. They were designed by people whose training came through apprenticeship and tradition rather than through university curricula, and whose mathematical tools were geometry, arithmetic, and a deep intuition for how materials behave when stacked, stretched, or submerged.

(Giza pyramids during the 19th century, having been built over 85 years between 2589 and 2504 BC.)

The Romans in particular deserve attention as the first large-scale civil engineers in the modern sense. Roman engineers built roads that connected a continent-spanning empire, aqueducts that delivered water over distances of a hundred kilometers with gradients measured in fractions of a percent, sewer systems that still partly function today, and harbor structures that used a specific volcanic concrete recipe more durable than most modern mixes. Much of what Roman engineers knew was lost or neglected in Western Europe for nearly a thousand years after the empire fell, though traditions continued in Byzantine, Islamic, Chinese, and Indian engineering through what Europeans called the medieval period.

(Roman aqueducts were first built in 312 BCE with the construction of the Aqua Appia. Construction continued for centuries, including major projects like the Aqueduct of Valens in the 4th century CE. Many aqueducts remained in use long after the fall of the Roman Empire in 476 CE, with some repaired and reused through the Middle Ages and beyond. Remarkably, structures such as the Segovia Aqueduct continued supplying water into the 20th century, demonstrating the durability and long-term functionality of Roman engineering.)

The phrase "civil engineer" emerged in the eighteenth century to distinguish nonmilitary engineering from military engineering, which until then had monopolized the word.

John Smeaton, an English engineer active in the mid-1700s, is generally credited as the first person to call himself a civil engineer. Smeaton's work on the third Eddystone Lighthouse, completed in 1759, illustrated the new posture of the profession. He studied the hydraulic behavior of lime mortars, tested materials systematically, and wrote detailed technical reports. His lighthouse stood for 120 years. In 1771 he helped found the Smeatonian Society of Civil Engineers, the first professional body dedicated to the field. The **Institution of Civil Engineers** followed in London in 1818, with Thomas Telford as its first president, and quickly became the model for similar bodies worldwide.

(An often overlooked detail is that John Smeaton conducted early experimental work on waterwheels and significantly improved their efficiency, making his designs widely adopted across 18th-century mills and industrial sites.)

The nineteenth century transformed civil engineering from a craft into an organized profession with formal education behind it. The École Polytechnique in Paris, founded in 1794, trained engineers using a rigorous mathematical and scientific curriculum that influenced the rest of Europe. In the United States, **Rensselaer Polytechnic Institute** in Troy, New York, offered the first civil engineering degree in the Western Hemisphere in 1835. West Point had been producing military engineers since 1802, and many of them went on to design civilian infrastructure. The railroad boom that followed, on both sides of the Atlantic, created demand for engineers who could survey routes through unfamiliar terrain, design cuts and embankments, and build bridges at a pace and scale no one had attempted before. By mid-century, the profession had specialties within it, codes of practice, technical journals, and enough collective knowledge that young engineers could inherit a body of work rather than reinvent fundamentals on every project.

Two advances in the late nineteenth and early twentieth centuries reshaped what civil engineers could build.

The first was the commercial availability of **structural steel**, produced by the Bessemer and later open-hearth processes, which offered tensile strength an order of magnitude beyond cast or wrought iron at a price that allowed it to replace those materials.

The second was the refinement of **Portland cement** and reinforced concrete, which let engineers combine concrete's compressive strength with steel's tensile strength in a single composite material that could be cast into almost any shape.

These two materials, more than any others, built the twentieth century. Skyscrapers, long-span bridges, hydroelectric dams, interstate highways, and the continuous-ribbon cities of the modern world are unthinkable without them.

The postwar decades saw civil engineering specialize further as projects grew in scale and complexity. Geotechnical engineering emerged as a distinct field after Karl Terzaghi's publication of Erdbaumechanik in 1925 established soil mechanics on a theoretical footing. Environmental engineering separated from sanitary engineering as pollution became a central public concern. Transportation planning absorbed tools from economics and operations research. By the late twentieth century, a civil engineer was expected to choose a specialization by the end of university and to keep learning within it throughout a career, while still maintaining enough breadth to coordinate with the other specializations on any real project.

The twenty-first century has pushed the profession in directions the twentieth didn't anticipate. Climate change, aging infrastructure in the wealthy world, rapid urbanization in the developing world, and the digital transformation of design and construction are all reshaping what civil engineers do. What hasn't changed is the core posture of the work, which is to take public responsibility for physical systems that serve large numbers of people over long periods of time. That responsibility is continuous from Smeaton to the present and is the thread that ties the whole history together.

Branches and Specializations Within the Discipline

Civil engineering today is wide enough that no practitioner covers all of it. Most professionals work within one or two specializations and collaborate with others for the rest. Understanding the branches is important for two reasons: it clarifies what a civil engineer actually does on a given day, and it shows how the later chapters of this book map onto the profession's internal divisions.

Structural engineering is what most people picture when they hear the word civil. Structural engineers design buildings, bridges, towers, and any other framework that must carry loads safely. They work in concrete, steel, timber, masonry, and increasingly in hybrid and composite systems. Their central question is whether a proposed structure has enough strength, stiffness, and stability to resist everything that might ever act on it, including its own weight, the people and objects inside it, wind, earthquakes, and temperature changes.

Geotechnical engineering is concerned with the ground. Soil and rock are the foundation of every structure, and their behavior is heterogeneous, time-dependent, and often counterintuitive. Geotechnical engineers investigate sites, classify soils, measure their strength and deformation properties, and design foundations, retaining walls, embankments, and slopes. They're also the first called when something settles, slides, or subsides. They must account for groundwater flow, pore pressure, and consolidation effects that evolve over time and influence stability. Modern practice also relies heavily on in-situ testing and numerical modeling to predict performance under complex loading conditions.

Transportation engineering covers the networks that move people and goods: roads, highways, railways, airports, ports, and increasingly the logistics of urban freight and micromobility. Transportation engineers design the physical infrastructure, analyze traffic flow, and plan how networks evolve over time. Their work mixes geometric design, pavement engineering, and a strong dose of behavioral analysis. They use data-driven models to optimize capacity, safety, and travel time across entire networks. Increasingly, they also integrate intelligent transportation systems and real-time data to manage congestion and improve system resilience.

Water resources and hydraulic engineering deals with water in motion and at rest: rivers, reservoirs, canals, storm drains, pipe networks, and treatment plants. Hydraulic engineers apply fluid mechanics to practical systems, from sizing a culvert under a rural road to managing a regional watershed.

Environmental engineering grew out of sanitary engineering and focuses on protecting human and ecological health. This includes drinking water treatment, wastewater treatment, solid and hazardous waste management, air quality, and remediation of contaminated sites. Environmental engineers work at the interface of chemistry, biology, and civil infrastructure.

Construction engineering and management is the field that turns designs into built reality. It covers estimating, scheduling, procurement, site logistics, and safety. Construction managers coordinate between designers, owners, contractors, subcontractors, and regulators, and their work determines whether a project is delivered on time, on budget, and as intended.

Surveying and geospatial engineering is the measurement arm of the profession. Surveyors establish the horizontal and vertical reference systems every project depends on, stake out construction work, and document as-built conditions. The field has been transformed by GPS, laser scanning, and drone-based photogrammetry.

Urban and municipal engineering focuses on the shared infrastructure of towns and cities: streets, sidewalks, sewers, water mains, streetlights, and the coordination between all of them. It sits close to urban planning and is often practiced inside municipal public works departments. Engineers in this field manage aging infrastructure, maintenance cycles, and capital improvement programs under tight budget constraints. They also coordinate utility conflicts and phased construction to minimize disruption in dense, continuously operating urban environments.

Other recognized branches include **coastal and ocean engineering**, which addresses shorelines, harbors, and offshore structures; **earthquake engineering**, which cuts across structural and geotechnical work in seismic regions; **forensic engineering**, which investigates failures and legal disputes; and **materials engineering**, which overlaps heavily with civil but is often treated separately. Sustainability, resilience, and digital engineering are less standalone branches than cross-cutting concerns that every specialization is absorbing.

The boundaries between specializations aren't rigid. A foundation design is both structural and geotechnical. A highway project involves transportation, pavement, drainage, and structural engineers. A water treatment plant needs process engineers, structural engineers for the tanks and buildings, and geotechnical engineers for the foundations. One of the skills a civil engineer develops over a career is knowing which questions belong in which specialization and when to bring others into the conversation.

Role of the Civil Engineer in Society

Civil engineers occupy an unusual position between public and private interest. Their clients are usually private owners or public agencies, but their designs affect people who had no say in the contract and who often can't even identify the engineer by name. A badly designed bridge endangers everyone who crosses it. A poorly sited treatment plant affects a community's health for decades. This structural feature of the work, that outcomes land on third parties, is why every mature jurisdiction in the world regulates civil engineering practice and why the profession's own ethical codes elevate public welfare above client loyalty.

In practical terms, civil engineers act as translators between three groups: the people who want something built, the people who will build it, and the public that will live with it afterward.

The first group, owners and developers, speak in terms of program, schedule, and budget.

The second group, contractors and trades, speak in terms of means, methods, and sequencing.

The third group, the public, speaks through regulation, zoning, political process, and sometimes litigation.

The engineer has to speak all three languages and hold them in productive tension. Designs that ignore owner economics don't get built. Designs that ignore constructability don't get built correctly. Designs that ignore public welfare don't deserve to be built.

The profession also serves a longer time horizon than most. Infrastructure has design lives measured in decades, and the most successful projects serve far longer than their nominal design life. The Brooklyn Bridge opened in 1883 and is still in daily use. The Pantheon in Rome still has the largest unreinforced concrete dome in the world, nearly two thousand years after it was built. Civil engineers plan, explicitly and routinely, for users who haven't been born yet. That time horizon shapes how the profession thinks about risk, maintenance, adaptability, and material choice.

Civil engineers also carry, collectively, much of the responsibility for how societies respond to long-term environmental and demographic change. Decisions about where to route a highway, whether to build levees or restore wetlands, how to power a water treatment plant, or whether to reinforce an existing building rather than replace it, aggregate into the trajectory of a region's carbon emissions, resilience to climate shocks, and quality of life for its residents. The profession has grown more conscious of this responsibility over the last generation, and the tools it uses to evaluate projects, from life cycle assessment to climate-adjusted design codes, reflect that shift.

Professional Ethics, Licensure, and Codes of Practice

The public-facing consequences of civil engineering work are why the profession is regulated more tightly than almost any other field of engineering. In most countries, certain categories of design work can be performed only by a licensed engineer, and the engineer's license is tied to personal accountability that extends for the life of the structure.

In the United States, licensure follows a four-step pathway. An aspiring engineer earns a degree from an **ABET-accredited** engineering program, typically a four-year Bachelor of Science in civil or a closely related engineering discipline. Near the end of their degree, they take the **Fundamentals of Engineering (FE)** exam, administered by the National Council of Examiners for Engineering and Surveying (NCEES). Passing the FE makes the candidate an Engineer in Training (EIT) or Engineer Intern (EI), depending on the state. They then complete a period of supervised practice, typically four years, under a licensed engineer. Finally, they take the **Principles and Practice of Engineering (PE)** exam in their chosen discipline, which for civil engineers is one of several civil-specific exams covering structural, geotechnical, transportation, water resources, or construction. Passing the PE and meeting all other state requirements confers the **Professional Engineer** license.

A Professional Engineer can sign and seal drawings and reports, and by doing so, takes personal legal responsibility for the work. The seal is not a formality. It is the public's guarantee that a competent, identified, accountable person has reviewed the design and stands behind it. Practicing engineering without a license, or signing work one isn't competent to review, can result in loss of license, civil liability, and in extreme cases criminal charges.

Other countries follow broadly similar pathways with different names. The United Kingdom uses the **Chartered Engineer (CEng)** designation, awarded by the Engineering Council through professional institutions including the Institution of Civil Engineers. Canada uses the **P.Eng** designation administered province by province. Australia uses Chartered Professional Engineer (CPEng). Many countries have reciprocity or mutual recognition agreements that let licensed engineers practice across borders with additional steps.

Licensure establishes competence. Ethics codes establish obligation. The **American Society of Civil Engineers (ASCE) Code of Ethics**, in its current form, opens with a foundational principle that engineers shall hold paramount the safety, health, and welfare of the public. That ordering is not accidental. When obligations conflict, public welfare comes first, before obligations to employers, clients, or even one's own economic interest. Similar principles appear in the National Society of Professional Engineers (NSPE) code and in the codes of every national engineering body worldwide.

Ethical conduct in civil engineering plays out in quieter ways than dramatic refusals to sign unsafe designs, though those moments do occur and make the news when they do. More commonly, ethics shows up in daily choices: maintaining competence by continuing education, disclosing conflicts of interest, not taking on work outside one's expertise, being honest in reports and estimates, and giving credit to collaborators. It also shows up in how engineers respond when they notice problems in others' work. Ethics codes generally require reporting of observed unsafe conditions or incompetent practice, not as a matter of preference but as an obligation of the license.

Codes of practice are the technical backbone of engineering regulation. These are detailed, often hundreds of pages long, documents that prescribe how specific types of work must be designed, constructed, and inspected. In the United States, the **International Building Code (IBC)**, adopted by most states, references a suite of material-specific standards: ACI 318 for reinforced concrete, AISC 360 for structural steel, NDS for wood, TMS 402 for masonry, ASCE 7 for loads. Transportation engineers follow the AASHTO standards. Water and wastewater engineers follow AWWA and Ten States Standards. Every subdiscipline has its code lineage, and part of professional competence is knowing which code applies to a given project, which edition of it governs, and how local amendments modify the base document.

Codes are not infallible. They represent accumulated professional judgment and are revised on a cycle, typically every three to six years. Engineers are expected to apply codes with understanding, not as cookbooks. A code-compliant design is a necessary condition for professional practice but is not always sufficient for a good design, and engineers sometimes go beyond code minimums when context warrants.

Units, Measurement Systems, and Engineering Notation

Numbers without units are meaningless in engineering. Miscommunicating units has caused spectacular failures, the most famous being the loss of the Mars Climate Orbiter in 1999 due to a mix of metric and imperial thrust values. Civil engineering practice worldwide is standardized on the **International System of Units (SI)**, also called the metric system, but the United States continues to use U.S. customary units, often called English or imperial units, for most civil work. American engineers routinely convert between the two and maintain familiarity with both.

The SI base units relevant to civil engineering are the meter (m) for length, the kilogram (kg) for mass, the second (s) for time, and the kelvin (K) for temperature. Derived units follow from these. Force is measured in newtons (N), where one newton equals one kilogram-meter per second squared. Pressure and stress are measured in pascals (Pa), where one pascal equals one newton per square meter. Because the pascal is small for most engineering purposes, kilopascals (kPa) and megapascals (MPa) are standard. Work and energy are measured in joules (J), and power in watts (W).

U.S. customary units for civil engineering use the foot (ft) for length, the pound-mass (lbm) or slug for mass, the second for time, and degrees Fahrenheit for temperature. Force is measured in pounds-force (lbf), and pressure in pounds per square inch (psi) or pounds per square foot (psf). The pound-force and pound-mass are distinguished by context: a pound-mass is a unit of mass, while a pound-force is the weight of a pound-mass under standard gravity. Confusing the two is a classic source of error. Stress is commonly expressed in psi or ksi (kips per square inch, where one kip equals 1,000 pounds-force).

The following table summarizes conversions between common civil engineering units.

Quantity	**SI Unit**	**U.S. Customary Unit**	**Approximate Conversion**
Length	meter (m)	foot (ft)	1 m = 3.281 ft
Length	millimeter (mm)	inch (in)	25.4 mm = 1 in exactly
Area	square meter (m^2)	square foot (ft^2)	1 m^2 = 10.76 ft^2
Volume	cubic meter (m^3)	cubic yard (yd^3)	1 m^3 = 1.308 yd^3
Mass	kilogram (kg)	pound-mass (lbm)	1 kg = 2.205 lbm
Force	kilonewton (kN)	kip (1,000 lbf)	1 kN = 0.2248 kip
Pressure, Stress	megapascal (MPa)	pound per square inch (psi)	1 MPa = 145.0 psi
Pressure, Stress	kilopascal (kPa)	pound per square foot (psf)	1 kPa = 20.89 psf
Unit Weight	kilonewton per cubic meter (kN/m^3)	pound per cubic foot (pcf)	1 kN/m^3 = 6.366 pcf
Moment	kilonewton-meter (kN·m)	kip-foot (kip-ft)	1 kN·m = 0.7376 kip-ft

Engineering notation expresses numbers as a coefficient multiplied by ten raised to a power that is a multiple of three. So 12,500 becomes 12.5×10^3, not 1.25×10^4. This convention aligns with SI prefixes (kilo-, mega-, giga-, milli-, micro-) and makes values easier to read at a glance. When writing engineering quantities, units always accompany numbers, and the number of significant figures reflects the precision of the underlying measurement or calculation. Reporting a soil cohesion as 23.4817 kPa when the test accuracy is closer to 5 kPa is misleading and unprofessional; the appropriate report is 23 kPa or perhaps 25 kPa.

Civil engineering also uses dimensional reasoning as a quality check. Any equation must balance dimensionally, meaning the units on both sides must be the same. An engineer who finds that a calculation has units of meters-squared on one side and meters-cubed on the other has found an error in the algebra, not an exotic result.

The Design-Build-Operate Lifecycle of Infrastructure

Every piece of civil infrastructure moves through a recognizable lifecycle, and understanding this lifecycle is how students begin to think like practitioners. The specific names of phases vary by country and agency, but the essential sequence is consistent: a project is conceived, planned, designed, procured, constructed, commissioned, operated and maintained, and eventually renewed or retired. Civil engineers participate at every phase, though the particular specialists involved shift as the project evolves.

The **planning phase** begins when an owner, public or private, identifies a need. This might be a deteriorating bridge that requires replacement, a new highway to relieve congestion, a water treatment plant to serve a growing population, or a commercial development on a previously undeveloped parcel. Planners and engineers study alternatives, estimate costs and benefits, evaluate environmental and social impacts, and help the owner decide whether and how to proceed. The deliverables of this phase are typically feasibility studies, environmental assessments, and conceptual designs that establish scope, budget, and schedule at a rough order of magnitude. Decisions made here, about site location, scale, and basic configuration, have more influence on the project's lifetime cost and performance than any subsequent decision.

The **preliminary design phase** refines the chosen concept to the point where its feasibility is confirmed and major technical issues are identified. Site investigation, including geotechnical borings, surveys, and hydraulic studies, is done here. Design alternatives are compared. The deliverable is a preliminary design report, often at 30 percent completeness, that establishes the basis of design and supports approvals, permits, and funding decisions.

The **final design phase** produces the documents that will be used to build the project: complete construction drawings, technical specifications, engineering calculations, and contract documents. Drawings show the geometry of every element; specifications describe the materials, workmanship, and performance requirements; calculations demonstrate that the design satisfies code and project requirements. In most jurisdictions a licensed professional engineer seals these documents before they can be used for permitting or construction, taking legal responsibility for their correctness.

The **procurement phase** brings a contractor onto the project. Different delivery methods exist. In **design-bid-build**, the traditional method, the owner completes design and then solicits competitive bids from contractors, awarding typically to the lowest responsive bidder. In **design-build**, a single entity takes responsibility for both design and construction, allowing earlier cost certainty and faster schedules at the cost of some owner control. In **construction manager at risk (CMAR)** and other collaborative delivery methods, the contractor is engaged during design to advise on constructability and cost, then commits to a guaranteed maximum price before construction begins. Public-private partnerships (**P3** or **PPP**) bring private capital into public infrastructure, with the private partner often handling some combination of design, construction, financing, and operation for a defined concession period. Each delivery method shifts risk and responsibility differently among the parties. Chapter 9 covers these arrangements in detail.

The **construction phase** translates drawings into physical reality. A general contractor mobilizes to the site, coordinates subcontractors for each specialty trade, manages materials, equipment, and labor, and executes the work under the terms of the contract. The design engineer remains involved during construction to answer questions, respond to unforeseen conditions, review submittals (shop drawings, material samples, and fabrication details), and observe the work for conformance to the design. Civil construction is weather-dependent, logistically complex, and sensitive to sequencing, so the construction phase rewards project management discipline as much as it rewards technical skill.

The **commissioning phase** verifies that the completed facility works as designed. For simple projects this may be a walk-through punch list; for complex ones like treatment plants, it involves systematic testing of every subsystem, operator training, and a gradual transfer of responsibility from contractor to owner. Commissioning closes the gap between what was built and what will actually be operated.

The **operation and maintenance (O&M) phase** is where the project spends most of its life. A bridge designed for a 75-year service life spends perhaps three years in design and construction and the remaining 72 in service. Maintenance is the quiet work that determines whether the design life is actually achieved. Inspection programs identify deterioration before it becomes dangerous. Routine maintenance (sealing pavement cracks, cleaning culverts, repainting steel) extends life. Rehabilitation (resurfacing, component replacement, strengthening) addresses larger problems. Engineers who design infrastructure increasingly consider maintainability as a design criterion, locating access points, selecting materials with longer service life, and documenting the design in ways that help future maintenance engineers make good decisions.

The **renewal or retirement phase** closes the cycle. At some point, a facility reaches the end of its useful life, either because physical deterioration makes continued operation uneconomic, because needs have changed and the facility no longer serves them, or because better alternatives exist. The facility may then be rehabilitated, repurposed, or decommissioned. Decommissioning itself is an engineering project, with its own design, permitting, and construction phases, and for certain facilities (nuclear plants, contaminated industrial sites, old dams) it can be more technically challenging than the original construction.

A defining feature of the lifecycle is that cost commitments and flexibility move in opposite directions. In the planning phase, very little money has been spent but almost every important decision is still open. By the time construction is underway, most of the money is being spent but most of the decisions are locked in. Changing a highway alignment during planning costs almost nothing; changing it during construction costs millions. The graph of cumulative decisions made against cumulative money spent is one of the most important mental models in civil engineering. It's why front-end planning gets disproportionate attention and why experienced engineers push back when asked to make fundamental changes late in a project.

Lifecycle thinking has expanded in recent decades to include environmental and social dimensions. **Life cycle assessment (LCA)** evaluates the total environmental impact of a facility from raw material extraction through end-of-life disposal, including the embodied carbon in construction materials and the operational emissions over decades of use. Chapter 12 returns to LCA and to the **digital twin** approach, in which a continuously updated virtual model of a facility tracks its state across the lifecycle, supporting maintenance, retrofit, and eventual retirement decisions with richer data than any previous generation of infrastructure enjoyed.

Every chapter that follows will connect back to this lifecycle. When Chapter 3 discusses concrete strength development, keep in mind that curing happens during construction, that serviceability governs the operation phase, and that durability determines when renewal becomes necessary. When Chapter 5 covers foundation design, remember that soil investigation happens in preliminary design, that construction tolerance is set in final design, and that settlement monitoring extends into operation. The lifecycle is the spine of civil engineering practice, and every technical subject the profession studies can be understood as serving one or more of its phases.

> **If civil engineers design systems that outlive them by decades, how should that time horizon change the way they make decisions today under uncertainty?**

Chapter 2: Engineering Mechanics and Statics

Every structure a civil engineer designs, from a backyard retaining wall to a suspension bridge spanning a mile of open water, stands still because the forces acting on it balance out. Statics is the study of that balance. Before an engineer can analyze how a beam deflects, how a column buckles, or how a foundation settles, she has to know precisely what forces are acting where and how they resolve.

This chapter covers the mechanical vocabulary that every other specialty in the profession borrows from: vectors, equilibrium, moments, reactions, and the geometric properties of cross sections that determine how efficiently a shape resists load.

Mechanics divides broadly into statics and dynamics. Dynamics deals with motion and the forces that cause it. Statics deals with bodies at rest, or moving at constant velocity, where the net force and net moment equal zero. Almost all civil engineering design lives in the static domain. A finished bridge doesn't accelerate down the road. A completed building doesn't translate or rotate. Even structures subjected to dynamic loads, earthquakes, wind gusts, traffic impacts, are typically designed using equivalent static forces that produce the same worst-case effect. This makes statics the bedrock of the discipline.

Force Vectors, Equilibrium, and Free-Body Diagrams

A **force** is a vector. That word matters. A vector has both magnitude and direction, and forces can't be added or analyzed without respecting both. A 500-pound load pushing straight down on a column is a fundamentally different problem from a 500-pound load pushing at 30 degrees off vertical, even though the numerical magnitude is identical. Ignoring direction is the single most common error in introductory statics work.

In two dimensions, a force is usually resolved into its horizontal and vertical components, typically called Fx and Fy. If a force F acts at angle theta from the horizontal, its components are Fx = F cos(theta) and Fy = F sin(theta). In three dimensions, the force has three components aligned with the x, y, and z axes. Engineers switch fluidly between describing a force by its magnitude and direction and describing it by its components, because different problems make one representation more tractable than the other. A rope pulling at a known angle is described most naturally by magnitude and angle. Forces meeting at a joint get added most easily in component form.

Addition of force vectors follows the parallelogram rule, or equivalently, the head-to-tail method. Two forces acting on the same point combine into a single resultant whose components are the algebraic sum of the original components. If three forces meet at a point and their resultant is zero, the point is in equilibrium, and the three forces form a closed triangle when drawn head to tail. That geometric fact is the basis for solving countless practical problems, including cable tensions, truss joints, and cantilever attachments.

The principle of **equilibrium** is stated formally by two conditions in two-dimensional problems and three in three-dimensional problems. In 2D, the sum of horizontal forces equals zero, the sum of vertical forces equals zero, and the sum of moments about any point equals zero. Symbolically, $\Sigma Fx = 0$, $\Sigma Fy = 0$, $\Sigma M = 0$. These three equations constrain the system. Any problem with exactly three unknowns, typically three support reactions or three internal forces, can be solved by writing and solving these three equations. Problems with more unknowns than equations are called statically indeterminate and require additional compatibility conditions drawn from the deformation of the structure.

The tool that makes equilibrium solvable in practice is the **free-body diagram**. A free-body diagram isolates a single object, or a piece of an object, from everything around it. Every external force acting on that isolated body is drawn as an arrow, labeled with its magnitude if known, and placed at its correct point of application. Supports get replaced by the reactions they provide. Connections get replaced by the forces they transmit. Gravity gets drawn as a downward arrow at the center of gravity of the body. Nothing is left implicit. If a force isn't on the diagram, it isn't in the equations, and the answer will be wrong.

Learning to draw a correct free-body diagram is the single most important skill in the first year of an engineering curriculum. Students who can draw them consistently go on to succeed in structural analysis, dynamics, and fluid mechanics. Students who can't struggle for the rest of their careers. The diagram forces the analyst to decide, explicitly, what's being isolated, what's pushing on it, and in which direction each force acts. This sounds procedural, but it's where most analytical errors originate, and it's where most of them get caught.

Some forces are easier to identify than others. External loads such as gravity, applied wind pressures, and vehicle weights are usually given in the problem statement. Reaction forces at supports must be inferred from the support type. A **pin support** resists horizontal and vertical translation but allows rotation, providing two reaction components. A **roller support** resists translation in only one direction, typically vertical, providing a single reaction. A **fixed support** resists translation in both directions and also resists rotation, providing three reactions: horizontal force, vertical force, and a moment. Knowing which reactions to draw for which supports is foundational, and the symbol for each support type is standardized across textbooks and codes so that drawings can be read quickly and unambiguously.

Contact forces introduce another subtlety. When one body rests on another, the surface transmits a **normal force** perpendicular to the contact plane, and potentially a **friction force** tangent to it. The normal force is always compressive, never tensile, unless the surfaces are bonded. Friction opposes impending motion and

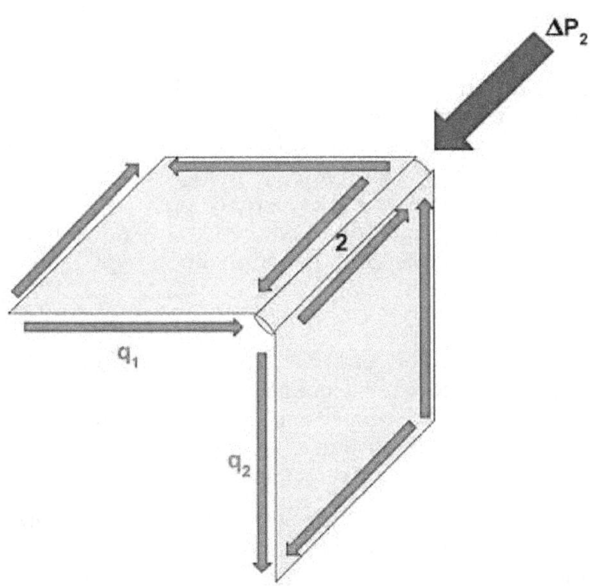

(This free-body diagram isolates a thin-walled structural corner and shows **distributed shear flows** (q_1, q_2) circulating along each panel due to an applied load ΔP_2. Each of the 8 arrows on the structure is a **force per unit length**, not a single force, so direction matters. These flows must be resolved and summed vectorially to satisfy equilibrium. The corner transfer (labeled "2") shows how forces change direction at joints, similar to head-to-tail vector addition. If all shear flows balance ($\Sigma F = 0$, $\Sigma M = 0$), the section remains in equilibrium.)

is limited by the coefficient of static friction multiplied by the normal force. Engineers working on retaining walls, slope stability, and heavy equipment anchorage lean heavily on these two quantities.

Internal forces are revealed by cutting through a member and exposing what's transmitted across the cut. At any internal section of a loaded beam, three internal forces act: an **axial force** along the member's length, a **shear force** perpendicular to its length, and a **bending moment** that tries to rotate the cut section. Engineers trace these internal forces along the length of every member to find where they peak, because those peaks govern the design. The exercise of cutting, exposing, and solving is the bridge between external equilibrium and the internal stresses that material behavior must resist.

Moments, Couples, and Reactions at Supports

A **moment** is the rotational effect of a force. If a force doesn't pass through a reference point, it tends to rotate bodies about that point. The magnitude of the moment equals the force multiplied by the perpendicular distance from the reference point to the force's line of action. This perpendicular distance is called the **moment arm** or **lever arm**. The further a force acts from a point, the larger its rotational effect, which is why a wrench with a longer handle loosens a stuck nut more easily.

Moments follow a sign convention, typically counterclockwise positive and clockwise negative, though the opposite convention appears in some texts. Consistency within a problem matters more than the choice. When several forces act on a body, each contributes a moment about any chosen reference point, and the total moment is the algebraic sum. The clever engineer picks the reference point to eliminate as many unknown forces as possible. Choosing a point where two unknown reactions pass through, for example, leaves only one unknown in the moment equation, making the problem direct.

A **couple** is a pair of equal, opposite, and parallel forces separated by a perpendicular distance. The net force of a couple is zero, which means the body experiences no translation, but it does experience a pure rotational tendency whose magnitude equals one force times the distance between them.

Couples appear often in civil engineering: the steering of a car, the torsion in a drive shaft, and importantly for structures, the internal resisting moment inside a bending beam. A couple has the useful property of being a "free vector" in planar problems, meaning its rotational effect on the body is the same regardless of where it's applied.

Support reactions complete the cast of forces on any structure. The support conditions an engineer specifies (pinned, roller, fixed, guided, or a spring) determine how many reaction components each support can develop and therefore how the structure responds. A simply supported beam has one pin and one roller, giving three reactions total, and the three equations of equilibrium solve them. A cantilever has a single fixed support with three reactions, all at one end. A continuous beam spanning multiple supports has more reactions than equations, meaning it's indeterminate and requires compatibility analysis.

When engineers design a building, they pay close attention to the distinction between **lateral** and **gravity** load paths. Gravity loads travel straight down through beams, girders, columns, and finally into foundations. Lateral loads, wind pressure on a facade, earthquake shaking, impact from a collision, travel horizontally through floor diaphragms into shear walls or moment frames and then down to the foundation. Each path requires its own support logic, and each set of supports contributes reactions that satisfy equilibrium in its own direction.

Real supports never behave exactly like the idealized models in textbooks. A "pin" in steel construction is usually a bolted or welded connection that transmits some moment. A "roller" might be an elastomeric bearing pad that resists a bit of horizontal force through friction. A "fixed" base plate loosens slightly under cyclic loading. Engineers account for this by designing connections to behave within a tolerance of the idealized behavior, and by using conservative safety factors to absorb the uncertainty. This gap between the mathematical abstraction and the physical reality is where experience complements textbook knowledge.

Centroids, Moments of Inertia, and Distributed Loads

The geometry of a cross section controls how a structural member resists load. Two beams of identical material but different cross-sectional shapes carry vastly different loads before failing. The properties that capture this geometric effect, centroids and moments of inertia, are therefore indispensable.

The **centroid** of a cross section is its geometric center, the point at which, if the shape were uniformly dense, the shape would balance. For simple shapes such as rectangles, circles, and triangles, centroid locations are tabulated. For composite shapes made up of simpler parts, the centroid is found by taking the weighted average of the component centroids, each weighted by its area.

The formula $x_{centroid} = \Sigma(A_i \times x_i) / \Sigma A_i$ applies along each axis independently. This computation sits at the heart of analyzing I-beams, channels, T-sections, and any built-up member assembled from plates or angles.

The centroid matters because bending stress varies linearly across a cross section, with zero stress at the centroidal axis and maximum stress at the extreme fibers. Locating the centroid correctly is step one in any bending analysis. A small error in centroid location cascades into every downstream calculation.

The **moment of inertia**, sometimes called the **second moment of area**, quantifies a cross section's resistance to bending about a given axis. It equals the integral of the square of the distance from the axis times the differential area. Units are length to the fourth power: inches to the fourth, millimeters to the fourth. The further the material sits from the bending axis, the more moment of inertia it contributes, which is why I-beams concentrate their material in flanges far from the neutral axis. The web holds the flanges apart and carries shear, but the flanges do most of the bending work.

For a rectangular section with base b and height h bending about the horizontal centroidal axis, the moment of inertia is $bh^3/12$. That cubic term on the height is why beam depth matters so much more than beam width. Doubling the width doubles the moment of inertia. Doubling the depth multiplies it by eight. This single relationship shapes most of structural design.

The **parallel axis theorem** lets engineers compute moments of inertia about axes other than the centroidal axis. It states that the moment of inertia about any parallel axis equals the moment of inertia about the centroidal axis plus the area times the square of the distance between axes: $I = I_{centroidal} + A \times d^2$. For composite sections, each component contributes its own centroidal moment of inertia plus its area-distance-squared term, all summed. The parallel axis theorem turns what looks like an intractable integration into straightforward addition.

The closely related **radius of gyration**, r, equals the square root of the moment of inertia divided by the area: $r = \sqrt{I / A}$. It represents the distance from the axis at which the entire area could be concentrated to give the same moment of inertia. The radius of gyration shows up most often in column buckling analysis, where a column's slenderness ratio, length divided by radius of gyration, predicts whether it will buckle elastically or yield before buckling.

Distributed loads differ from concentrated point loads in that they spread along a length or across an area. A person standing on a bridge applies roughly a point load. Snow blanketing a roof applies a distributed load. Distributed loads have units of force per length for line loads, typically pounds per foot or kilonewtons per meter, and force per area for surface loads, typically pounds per square foot or kilopascals.

For equilibrium purposes, a distributed load can be replaced by a single equivalent concentrated force whose magnitude equals the total load, the area under the distribution, and whose line of action passes through the centroid of the distribution. A uniform distributed load of w over a length L has a resultant of wL acting at the midpoint. A triangular load peaking at w_max over length L has a resultant of (1/2) × w_max × L acting at one-third of the length from the peak. These equivalent forces make it possible to compute reactions and moments on loaded members using the same equilibrium equations that apply to concentrated loads.

The distinction between the true distributed load and its equivalent resultant matters when computing internal shears and moments along the length of the member. The resultant is valid only for overall equilibrium calculations, not for figuring out what's happening internally at each section. To construct accurate shear and bending moment diagrams along a beam, engineers work with the actual distribution, integrating as they move along the span.

That integration becomes the foundation of every beam design calculation in structural engineering, and it carries forward into structural analysis where the full behavior of framed systems unfolds.

> **When a structure fails, how often is the root cause a misunderstanding of forces rather than a lack of strength, and what does that imply about engineering practice?**

Chapter 3: Properties and Behavior of Construction Materials

Materials are the vocabulary of civil engineering. An engineer can master every equation in structural analysis, but without knowing how the chosen material actually behaves, the calculations don't produce reliable structures. Concrete cracks in tension long before it reaches its compressive strength. Steel yields, then strain-hardens, then necks, then ruptures, all governed by its chemistry and processing. Timber creeps under sustained load and responds differently parallel to its grain than perpendicular to it. Masonry carries compression beautifully but fails brittlely in tension. Each of these families of materials brings a distinct profile of strengths, weaknesses, costs, and behaviors, and the engineer's job is to match each application to the material whose properties best fit the demand.

This chapter surveys the main construction materials used in civil infrastructure today. Concrete and steel dominate modern practice and receive the deepest treatment. Timber, masonry, and composites each have specialized roles that warrant careful exploration. For each material, the focus stays on how it's made, how its properties emerge from its composition, and how those properties influence design decisions.

Concrete Composition, Curing, and Strength Development

Concrete is the most widely used construction material on Earth, by a wide margin. Roughly 30 billion tons are produced globally each year, more than steel, aluminum, plastic, and wood combined. Every highway, bridge, dam, building foundation, airport runway, and sewer pipe in modern infrastructure depends on it. Its ubiquity flows from three genuinely remarkable properties: it can be cast into almost any shape, it develops enormous compressive strength, and its raw materials are cheap and widely available.

At its simplest, concrete is a three-part system. **Cement** is the glue. **Aggregate**, both fine and coarse, is the filler. **Water** initiates the chemical reaction that binds everything together. The cement most commonly used is ordinary Portland cement, invented in England in 1824 and refined into the standardized product used today. Portland cement is manufactured by grinding limestone and clay, burning the mixture at about 1450 degrees Celsius in a rotary kiln, and then grinding the resulting clinker with a small amount of gypsum.

The fired clinker contains four primary compounds: tricalcium silicate (C3S), dicalcium silicate (C2S), tricalcium aluminate (C3A), and tetracalcium aluminoferrite (C4AF). These shorthand notations, familiar to every concrete technologist, correspond to specific calcium-rich minerals whose proportions determine how the cement will behave when mixed with water.

When water hits cement, a set of exothermic reactions called **hydration** begins. Tricalcium silicate reacts quickly, producing calcium silicate hydrate (C-S-H) and calcium hydroxide. Dicalcium silicate reacts more slowly, adding additional C-S-H over weeks and months. C-S-H is the crystalline gel responsible for almost all of concrete's strength. It grows, intertwines, and binds the aggregate together into a dense, hard composite. Calcium hydroxide is a byproduct that gives concrete its high alkalinity, a pH above 12.5, which is what protects embedded reinforcing steel from corrosion.

Hydration doesn't stop when the concrete feels hard. Strength development continues for months and even years, though the curve flattens quickly. Concrete typically reaches about 40 percent of its 28-day strength by day three, 70 percent by day seven, and 100 percent at day 28. The 28-day strength, usually called **f'c**, is the reference value used in design calculations. Standard concrete mixes target f'c values between 3,000 and 6,000 pounds per square inch, though high-strength mixes now routinely exceed 10,000 psi and ultra-high-performance concrete can reach 25,000 psi or more.

Water-cement ratio is the single most important variable in determining concrete strength. The ratio, written as w/c, is the mass of water divided by the mass of cement.

Lower ratios yield stronger, denser, more durable concrete. Ratios around 0.40 produce high-strength, low-permeability mixes suitable for structural applications. Ratios above 0.60 produce weaker, more permeable concrete acceptable only for non-structural fill. The physical reason is simple: only about 25 percent water by mass of cement is chemically required to hydrate all the cement. Any additional water creates voids when it eventually evaporates, and voids reduce strength and allow corrosive agents to penetrate.

Aggregate occupies 60 to 80 percent of concrete's volume and has its own engineering properties. Coarse aggregate is typically crushed stone or gravel, graded between 3/8 inch and 1.5 inches. Fine aggregate is natural or manufactured sand below 3/8 inch. The aggregate should be clean, well-graded so particles of different sizes pack densely, and strong enough that the aggregate itself isn't the weak link. Reactive aggregates can cause alkali-silica reaction, a slow chemical attack that expands the concrete from the inside and cracks it over decades. Engineers test aggregates before use to avoid this and other durability problems.

Modern concrete almost always contains **admixtures**, small quantities of chemicals or minerals that modify specific properties. Water-reducers let engineers achieve workable concrete at lower water-cement ratios, boosting strength without sacrificing placement ease. Superplasticizers, a more aggressive class of water-reducers, produce flowing, self-consolidating concrete that fills intricate forms without vibration. Retarders slow the setting time, which matters on hot days or long pours. Accelerators speed it up for cold-weather work. Air-entraining agents create microscopic bubbles throughout the mix that give the concrete freeze-thaw resistance: ice crystals can expand into the voids rather than cracking the matrix. Supplementary cementitious materials like fly ash, slag cement, and silica fume replace a portion of the Portland cement and contribute to long-term strength and durability while reducing the carbon footprint of the mix.

Curing is the process of keeping concrete moist and at a reasonable temperature while hydration proceeds. If concrete dries too quickly, hydration stops and final strength is reduced. If it freezes before it sets, the expanding ice destroys the developing matrix. Common curing methods include spraying with water, covering with wet burlap or plastic sheets, applying curing compounds that seal the surface, and using heated enclosures in cold weather. Seven days of moist curing at ordinary temperatures delivers most of the strength a concrete mix will ever develop. Fourteen days is better. Skipping curing is one of the most damaging shortcuts in construction, and its effects aren't visible until years later when surface crazing, scaling, or corrosion-related cracking appear.

Concrete's great weakness is tension. A typical concrete's tensile strength is about 10 percent of its compressive strength, and it fails in tension brittlely without warning. This fundamental asymmetry is why almost all structural concrete is **reinforced** with steel bars, prestressed with high-strength tendons, or both. Plain concrete works where loads stay in compression: gravity dams, mass foundations, and unreinforced walls. Everything else, including every beam, slab, and column subject to bending, requires steel reinforcement working in partnership with the concrete, with the concrete handling compression and the steel handling tension.

Structural Steel Grades and Metallurgical Properties

Steel is the other backbone of modern civil construction. It offers high strength in both tension and compression, predictable behavior under load, and the ability to be rolled, welded, bolted, and shaped into an extraordinary variety of profiles. Where concrete dominates by volume, steel dominates by performance per pound.

Structural steel is an iron-carbon alloy with tightly controlled additions of other elements. Plain carbon steels contain up to about 2 percent carbon, though structural grades stay well below 0.5 percent. Higher carbon content raises strength and hardness but reduces ductility and weldability. The sweet spot for structural shapes is low-carbon steel with carbon content between 0.15 and 0.25 percent, yielding a material that's strong, ductile, and readily welded.

The most common structural steel grade in North America is **ASTM A992**, the standard for wide-flange shapes used in buildings. It has a minimum yield strength of 50 ksi (kilopounds per square inch) and a minimum tensile strength of 65 ksi. **A572 Grade 50** covers plates and bars at the same strength level. **A36**, an older general-purpose grade with 36 ksi yield strength, still appears in angles, channels, and smaller members. **A500** and **A1085** cover hollow structural sections, the square and rectangular tubes used in trusses and exposed structures. High-performance **A709** grades are used in bridge construction, with weathering variants that form a protective oxide layer and eliminate the need for paint.

A **stress-strain curve** describes how steel responds to axial load. In the initial elastic range, strain is proportional to stress, following Hooke's law: stress equals modulus of elasticity times strain. The modulus of elasticity for structural steel is

29,000 ksi, remarkably consistent across all grades. This consistency is a gift to designers: the same value appears in every calculation, regardless of the specific steel being used.

Beyond the elastic limit, the steel yields. For mild structural steels, yielding occurs at a sharp point, the **yield stress**, after which strain increases while stress remains roughly constant through the yield plateau. Past the plateau, strain hardening begins: stress rises again as the internal crystal structure rearranges, eventually reaching the **ultimate tensile strength**. Beyond that peak, the cross section begins to neck, stress in terms of engineering stress drops, and the specimen ruptures. The area under the stress-strain curve up to rupture represents the total energy absorbed, and it's a measure of toughness. Ductility, the ratio of strain at rupture to strain at yield, is typically 100 to 200 for structural steel, which is why steel absorbs large amounts of energy before failure. This ductility is the reason steel structures survive earthquakes that would shatter brittle materials.

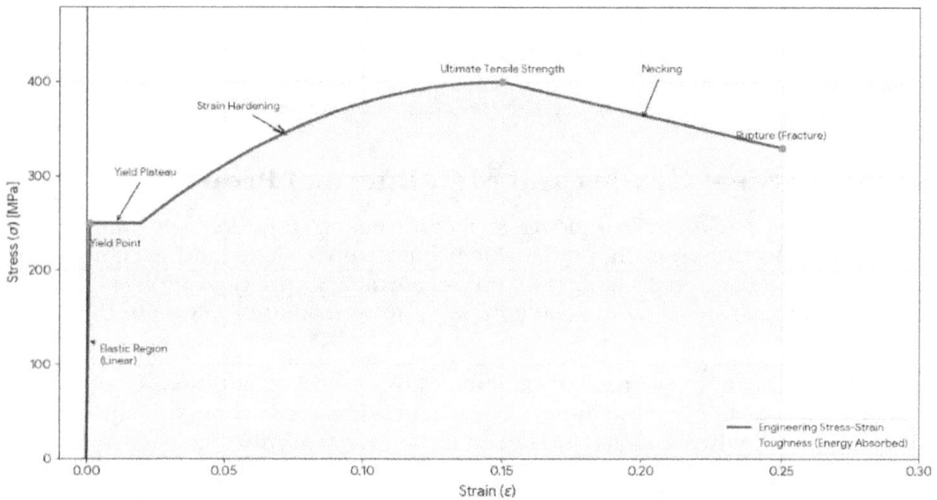

Temperature changes the picture. Steel's strength drops substantially above about 1000 degrees Fahrenheit, which is why unprotected steel loses most of its capacity in a building fire within minutes. Fireproofing, whether spray-applied cementitious coatings, intumescent paints, or gypsum board enclosures, extends the time the steel can maintain adequate strength and gives occupants time to evacuate. At the other extreme, steel becomes increasingly brittle as temperature drops below about -40 degrees Fahrenheit, and the grade must be chosen with cold-weather toughness in mind.

Weldability is a fundamental property for steel construction. Welds are the connections that tie members together, and a weld that fails is catastrophic. Carbon content, phosphorus, sulfur, and other residual elements influence how readily a steel can be welded without cracking. Carbon equivalent, a weighted sum of element percentages, predicts weldability: values below about 0.40 are considered readily weldable. Modern structural grades are intentionally formulated to weld well with standard processes such as shielded metal arc welding, flux-cored arc welding, and gas metal arc welding.

Corrosion is steel's persistent enemy. Atmospheric moisture, oxygen, chlorides from sea spray or de-icing salts, and carbon dioxide all attack exposed steel, slowly converting it to rust and reducing cross section. Protection strategies include paint systems, galvanizing (a zinc coating that sacrifices itself rather than the steel beneath), stainless steel grades in aggressive environments, and weathering steels that form a stable, self-sealing oxide layer. In concrete, steel is protected by the surrounding alkalinity, as long as the concrete cover is adequate and the concrete itself isn't carbonated by years of exposure to atmospheric CO2.

The shapes in which structural steel is rolled are standardized in the AISC Manual of Steel Construction. **Wide-flange shapes**, designated with the W prefix, are the workhorse: two flanges connected by a web, forming an I-shaped cross section. W14x90 denotes a wide flange about 14 inches deep weighing 90 pounds per foot. **Channels** (C) have one open side and show up as stair stringers and light framing. **Angles** (L) are workhorses in bracing and trusses. **Hollow structural sections** (HSS) include square, rectangular, and round tubes. Plates, bars, and pipes round out the catalog. Designers pick from this menu based on what moments and forces each member needs to resist and the connections it needs to accommodate.

Timber, Engineered Wood, and Mass Timber Products

Wood is the oldest structural material, and after a century of declining share in commercial construction, it's experiencing a renaissance. New engineered wood products have expanded what's possible, letting designers build mid-rise and even high-rise buildings in wood with competitive performance, faster construction, and lower embodied carbon than concrete or steel.

Natural wood, often called **solid-sawn lumber**, is harvested from trees and processed into rectangular cross sections at sawmills. Its properties vary with species, grain orientation, moisture content, and presence of defects such as knots, splits, and slope of grain.

Softwoods, conifers like Douglas fir, southern pine, spruce, pine, and fir, dominate structural applications because they grow quickly, yield long straight pieces, and are easily worked. Hardwoods, deciduous species like oak and maple, are stronger and heavier but more commonly used for flooring, furniture, and specialty work rather than structural framing.

Wood is **anisotropic**: its properties depend on the direction of loading relative to the grain. Strength and stiffness parallel to the grain are many times greater than those perpendicular to the grain. A spruce 2x10 loaded parallel to its length can carry thousands of pounds, but the same piece crushed across its width yields under much less load. Connections, bearing areas, and detailing all depend on grain direction, which is why building codes tabulate separate allowable stresses for tension parallel, tension perpendicular, compression parallel, compression perpendicular, bending, and shear.

Moisture content matters hugely. Wood shrinks as it dries below its fiber saturation point, around 28 to 30 percent moisture content, and swells as it absorbs moisture above that threshold. A framing member installed at 19 percent moisture and equilibrating in service at 12 percent can change dimensions measurably. Designers specify lumber at the moisture content closest to service conditions, typically kiln-dried to 19 percent or less for interior framing. Grade stamps on each piece identify species, grade, moisture content at processing, and the mill that produced it.

Timber design in North America uses the American Wood Council's National Design Specification for Wood Construction, which tabulates base values for each species and grade and provides adjustment factors for load duration, moisture, temperature, repetitive use, and size. Because wood creeps under sustained load and its strength depends on how long the load is applied, a piece of timber that can support a heavy snow load for a few weeks cannot support the same load as a permanent fixture.

Engineered wood products extend the capabilities of raw timber by manipulating wood at a smaller scale and reassembling it with adhesives. **Glued laminated timber**, or glulam, is made by gluing dimensional lumber boards together to form large beams and columns. Glulams can span enormous distances, dozens of feet in a single piece, and they can be curved to follow any architectural profile. **Laminated veneer lumber** (LVL) is made from thin wood veneers glued together with parallel grain, yielding high strength and dimensional stability. **Parallel strand lumber** (PSL) takes long strands of veneer, aligns them parallel, and bonds them under pressure.

Cross-laminated timber (CLT) is the breakthrough product behind the mass timber movement. CLT panels are made by stacking layers of dimensional lumber at right angles to each other and bonding them with structural adhesive, typically in odd numbers of layers: three, five, seven, or nine. The resulting panels act as two-way plates, competing with concrete slabs for floor and wall applications. CLT enables wood buildings up to 18 stories and higher under current codes, and installations go up in weeks rather than months because the panels are prefabricated, delivered to site, and craned into place.

Mass timber structures can be cost-competitive with concrete and steel in many markets, depending on local labor costs, fire-code acceptance, and supply chains. Their environmental profile is compelling: wood sequesters carbon during growth, and using it structurally locks that carbon up for the life of the building. Each cubic meter of CLT represents roughly a ton of $CO2$ kept out of the atmosphere. Fire performance, once the primary concern, is now well understood: mass timber elements char on the surface at predictable rates, and the char layer insulates the interior wood from further burning. Codes accept mass timber in larger buildings when the sections are sized so that the remaining cross section after a two-hour fire retains adequate structural capacity.

Masonry, Mortar, and Bonding Techniques

Masonry is construction assembled from individual units, bricks, concrete block, stone, glass block, bonded together with mortar. It's the oldest surviving structural system, dating back 10,000 years, and it remains common in residential foundations, low-rise walls, facades, chimneys, and many mid-rise building systems in regions where labor is inexpensive.

Clay brick is manufactured by forming clay into rectangular units, drying them, and firing them in kilns at temperatures between 1600 and 2000 degrees Fahrenheit. The firing vitrifies the clay and produces a unit with compressive strength ranging from 2,500 to over 10,000 psi. Standard modular brick measures approximately 3-5/8 by 2-1/4 by 7-5/8 inches, sized so that three courses of brick plus mortar joints equal 8 inches vertically. Variations in color, texture, and finish are produced by controlling the clay chemistry, firing conditions, and surface treatment. Concrete masonry units, commonly called concrete block, are cast in molds from a stiff concrete mix and cured. Standard 8-inch block measures nominally 8 by 8 by 16 inches, including the mortar joint, and is hollow to reduce weight and allow reinforcement and grout to be placed in the cells.

Mortar is the material that bonds individual units and transfers load between them. It's made from cement, lime, sand, and water, proportioned to balance strength, workability, and water retention. Types M, S, N, O, and K, in decreasing order of strength, are specified in ASTM C270. Type S mortar, with a minimum compressive strength of 1800 psi, is commonly used for structural walls. Type N, at 750 psi, is used for exterior non-load-bearing walls. Stronger mortar isn't always better: a mortar that's too hard transfers stress concentrations to the weaker masonry units and can crack them, while a softer mortar absorbs small movements and keeps the assembly intact.

Masonry's structural behavior is dominated by its behavior in compression. Individual brick and block units can carry thousands of psi in pure compression, and well-built masonry walls carry the dead weight and live loads of multi-story buildings easily. The weakness, again, is tension. Mortar joints transfer almost no tensile stress, so masonry walls crack readily under bending, lateral loads, and differential settlement. Modern masonry design resolves this by introducing steel reinforcement into the hollow cells of concrete block or between brick wythes, then filling the cells with grout to bond the steel to the masonry. This **reinforced masonry** can resist earthquake loads, wind loads, and tensile stresses in the same way as reinforced concrete.

Bond patterns affect both appearance and structural performance. **Running bond**, where each brick is offset half a unit from the course below, is the most common pattern and provides good load distribution by ensuring no vertical joint runs through more than one course. **Stack bond**, where units stack directly on top of each other in vertical columns, is weaker structurally because loads can concentrate along continuous vertical joints, and it typically requires additional horizontal reinforcement. English bond, Flemish bond, common bond, and other historic patterns appear in brickwork and have distinctive visual effects, each offering different combinations of strength and appearance.

Control joints accommodate thermal and moisture movement in masonry walls. Without them, temperature expansion and shrinkage of the materials would cause uncontrolled cracking. Control joints are typically spaced 20 to 30 feet apart horizontally and placed at locations where stress concentrations naturally occur: the corners of openings, changes in wall height, and intersections with dissimilar materials.

Composite Materials and Fiber-Reinforced Polymers

Composite materials combine two or more constituents whose combined performance exceeds what any one of them could deliver alone. Reinforced concrete is itself a composite in this sense, combining the compressive strength of concrete with the tensile strength of steel. But the term "composite" in modern engineering typically refers to **fiber-reinforced polymers** (FRP), advanced composites originally developed for aerospace that have found growing application in civil infrastructure over the past three decades.

An FRP is built from strong, stiff fibers embedded in a softer polymer matrix. The fibers carry the primary load. The matrix holds the fibers in place, distributes load between them, and protects them from the environment. Common fiber types include carbon, glass, aramid (Kevlar), and basalt. Each has a distinct profile.

Carbon fiber offers the highest strength-to-weight ratio and stiffness but costs far more than the alternatives.

(A less-known fact is that carbon fiber exhibits a **negative coefficient of thermal expansion** along the fiber direction, meaning it can slightly contract when heated, a property engineers exploit in precision structures where dimensional stability is critical.)

Glass fiber is much cheaper and still delivers impressive properties, making it the most widely used FRP reinforcement in civil applications. **Aramid fiber** excels at absorbing impact and appears in structures that must resist blast or ballistic loads.

Polymer matrices include epoxy, polyester, vinyl ester, and thermoplastic resins. Epoxy offers the best mechanical properties and environmental durability, and it's used in most structural-grade FRP products. Polyester and vinyl ester are cheaper and common in pipe, tanks, and non-structural applications. Matrix selection affects stiffness, long-term durability, moisture resistance, temperature capability, and cost.

In civil engineering, FRP appears in several distinct forms. **FRP bars** substitute for steel rebar in concrete applications where corrosion is a persistent threat: marine structures, bridge decks exposed to de-icing salts, parking garages, and concrete in contact with aggressive chemicals. Glass FRP bars are cost-competitive with epoxy-coated steel and don't rust at all, though they lack the ductility of steel and have lower modulus of elasticity, requiring different design methods. **FRP wraps** consist of fabric sheets saturated with epoxy and applied to the exterior of existing concrete members. Column wrapping adds confinement and dramatically increases compressive capacity. Beam wrapping adds flexural or shear strength where the original design is deficient. **FRP plates** bonded to the soffit of beams strengthen them in bending with minimal added weight.

The biggest growth area for FRP in civil infrastructure is seismic retrofit and strengthening of existing buildings and bridges. Thousands of concrete columns in older structures lack the transverse reinforcement required by modern codes to survive strong earthquakes. Wrapping these columns with carbon FRP adds the needed confinement quickly, cleanly, and without adding significant mass that would worsen the seismic demand.

Non-fiber-based composites also have a place. **Fiber-reinforced concrete** adds short fibers, steel, polypropylene, glass, or synthetic, to the concrete matrix, improving toughness, crack control, and post-cracking behavior. **Ultra-high-performance concrete** (UHPC) combines Portland cement, silica fume, fine sand, and fine steel fibers in a dense, low-water-cement-ratio mix that can reach compressive strengths above 25,000 psi and maintains tensile capacity through fiber bridging.

Composite systems face several constraints in civil applications. Durability under sustained load, called creep rupture, limits the allowable long-term stress in FRP bars to a fraction of their short-term strength. Sensitivity to ultraviolet light and moisture demands protective coatings or embedment. Lack of ductility means FRP-reinforced structures fail suddenly rather than warning with visible distress, and designers must compensate with conservative safety factors. High initial cost keeps FRP from replacing steel in routine new construction, though the life-cycle economics improve when corrosion would otherwise drive expensive replacements. These constraints are the reason composites remain specialty rather than commodity materials in civil engineering, even as their use grows year over year in niches where their unique combination of properties justifies the price.

If every material has fundamental weaknesses, how does an engineer decide which trade-offs are acceptable in a given design?

Chapter 4: Structural Analysis and Design

Structural analysis is the process of predicting, before anything is built, how a structure will respond to everything that will act on it over its service life. Design is the process of making sure that response stays within acceptable limits. The two go hand in hand: analysis asks "what will happen?" and design asks "is that safe enough?" Neither one is useful without the other.

This chapter connects the mechanics principles and material properties established earlier with the practical methods engineers use to size beams, proportion columns, shape trusses, and verify that a finished structure won't collapse, deflect excessively, vibrate unpleasantly, or deteriorate prematurely.

The discipline rests on three assumptions that become second nature to practicing engineers. The first is that the loads a structure will experience can be predicted with reasonable probability, not certainty, but enough statistical confidence to design around. The second is that the material will behave as the specifications say it will, within a known tolerance. The third is that the analytical model, however simplified, captures the dominant behavior of the real structure. Safety factors and load factors exist because none of these assumptions is ever exactly true.

Load Classification: Dead, Live, Wind, Seismic, and Thermal

Every structure carries loads, and the first task in design is identifying what those loads are, how large they might become, and how often each might occur. Building codes, notably ASCE 7 in the United States, the International Building Code, and Eurocode 1 in Europe, tabulate loads for standard occupancies and expose the methodology for non-standard conditions. Classifying loads correctly is not a trivial exercise. A mistake in load identification cascades through every downstream calculation and often isn't caught until a failure exposes it.

Dead load is the permanent self-weight of the structure and everything attached to it that doesn't move during normal use. The weight of the structural frame itself, the slab, the cladding, partition walls, permanent mechanical equipment, roofing, flooring, and fixed ceilings all fall into this category. Dead load is estimated from material densities and member sizes. Steel weighs about 490 pounds per cubic foot. Normal-weight concrete weighs about 150 pounds per cubic foot. Softwood lumber runs around 30 to 40 pounds per cubic foot depending on species and moisture. These densities are well known, so dead load is typically estimated with high confidence and accompanies the structure throughout its life without changing.

The catch is that dead load must be assumed before the members are sized, which creates a chicken-and-egg problem. Engineers solve it by estimating member sizes based on experience or preliminary calculations, computing the corresponding dead load, and iterating if the final sizes differ enough from the initial guesses to matter. Most iterations converge quickly, and modern software handles the update automatically.

Live load is the transient, movable load associated with occupancy and use. People, furniture, movable partitions, vehicles, stored inventory, and similar items all contribute. Live load is inherently variable: a classroom occupied with 30 people has more live load than the same classroom empty, and a library with floor-to-ceiling bookshelves has far more live load than a standard office. Code tabulates uniform live loads by occupancy. A typical office sees 50 pounds per square foot. Public assembly spaces see 100 psf. Library stack rooms see 150 psf. Warehouses can exceed 250 psf depending on the goods stored. Concentrated live loads, such as the weight of a single heavy piece of equipment, must be checked separately where they might govern the design of an individual member.

Live loads may be reduced on members carrying large tributary areas, because the probability that every square foot of a large floor will simultaneously carry its full design live load diminishes as the area grows. Code provides reduction formulas that engineers apply to beams, girders, and columns based on the area each supports. The reductions can be substantial, 40 percent or more for columns supporting many floors, and they reflect the statistics of occupancy rather than any reduction in the worst-case local load.

Snow load is a special case of a roof live load that depends on geography, roof geometry, exposure, and thermal condition of the roof. Ground snow loads tabulated in code serve as the starting point and get adjusted by importance factors, exposure coefficients, thermal coefficients, and slope factors. A steep, unheated roof in a protected location accumulates snow differently than a flat, heated roof in a windswept area. Drift loads from wind redistributing snow across roof elevations or against parapets can locally exceed the uniform snow load by large factors and must be checked on every roof with projections or changes in elevation.

Wind load is the pressure that moving air applies to the surfaces of a structure. It depends on wind speed, building geometry, exposure category, topography, and the degree of enclosure. Basic wind speed is mapped across the country in the design code, with values ranging from 90 miles per hour in sheltered areas to 180 miles per hour or higher in hurricane-prone coastal regions. The pressure on a surface is computed from the velocity pressure q, which is proportional to the square of wind speed, multiplied by coefficients that account for the surface orientation and its relationship to the building envelope.

Wind exerts different effects on different surfaces of a building simultaneously. The windward wall sees positive pressure pushing inward. The leeward wall, side walls, and most of the roof see negative pressure (suction) pulling outward. The combined effect generates both lateral loads that the structure must resist as a whole and local pressures that cladding, fasteners, and individual components must handle. Roof corners and edges experience especially high suction pressures because of airflow separation, and connections in these zones must be detailed to match.

Tall and slender structures face an additional concern: dynamic wind response. Wind gusts excite the structure dynamically, and structures with low natural frequencies can resonate, amplifying the response beyond what a static analysis would predict. Gust-effect factors in code account for this, and truly slender or tall structures often require wind tunnel testing to quantify the dynamic response accurately.

Seismic load is the inertial force generated when ground motion during an earthquake accelerates the mass of the structure. Unlike wind, which acts on surface area, seismic force is proportional to mass. Heavier structures experience larger earthquake forces. The magnitude depends on the expected ground motion at the site, the dynamic characteristics of the structure, the soil type, and the ductility of the structural system. Code procedures range from simple equivalent lateral force methods, where a single horizontal force is applied at each floor, to response spectrum analysis and full nonlinear time history analysis for complex or important structures.

The equivalent lateral force method starts from a design spectral acceleration determined by location and soil type, multiplies by the structure's weight, divides by a response modification factor R that accounts for the structure's ductility and redundancy, and distributes the resulting base shear over the height of the building.

The R factor is perhaps the most consequential single parameter in seismic design: a special moment frame might have R = 8, while an ordinary moment frame has R = 3, reflecting the different ductility each can reliably provide. Higher R means lower design force but stricter detailing requirements to ensure the assumed ductility is actually available when the earthquake arrives.

Thermal load is the internal stress a structure develops when temperature changes cause materials to expand or contract but the structure's connections restrain the movement. A steel beam heated 100 degrees Fahrenheit wants to expand about 0.0065 percent of its length. If the ends are restrained against movement, the beam develops a compressive stress of roughly 19,500 psi from the thermal effect alone, approaching yield for A36 steel. Long bridges, long buildings, and concrete slabs all face thermal issues, and expansion joints, bearing pads, and movement details accommodate the expansion without inducing destructive stresses. Concrete's coefficient of thermal expansion is similar enough to steel's that the composite material doesn't develop significant thermal stresses within itself under normal temperature swings, which is one of the under-appreciated reasons reinforced concrete works.

Other loads occasionally govern. Ice accumulation on transmission towers, radio masts, and exposed steel in freezing climates adds substantial weight and wind area. Hydrostatic pressure from groundwater loads basement walls and tanks. Earth pressure loads retaining walls and below-grade walls. Flood loads, tsunami loads, blast loads, and vehicular impact loads all have specific code procedures for structures exposed to those hazards. In each case, the engineer determines which loads can act simultaneously and combines them using prescribed load combinations.

Load combinations recognize that not every load reaches its maximum value at the same moment. The code combination 1.2D + 1.6L + 0.5(Lr or S or R) for ultimate strength, for instance, factors up the dead and live loads by different amounts and adds a reduced roof live, snow, or rain load. A separate combination, 1.2D + 1.0W + L + 0.5(Lr or S or R), governs when wind is acting with dead and live. An earthquake combination appears. A set of service-level combinations applies for deflection checks. Design must satisfy every applicable combination, and the governing combination often differs from member to member within the same structure.

Beam Theory, Shear, and Bending Moment Diagrams

A beam is a structural member loaded perpendicular to its longitudinal axis that resists those loads primarily through bending. Beams occupy more line-feet of any building than any other structural element, and their analysis is foundational to every structural specialty. The classical theory that governs beam behavior, variously called Euler-Bernoulli beam theory or engineering beam theory, rests on a few idealizations that yield remarkably accurate predictions for slender, elastic, homogeneous beams.

The first idealization is that plane sections remain plane and perpendicular to the neutral axis after bending. This geometric assumption means that strain varies linearly across the depth of the beam, from compression on one face to tension on the other, passing through zero at the neutral axis. For linear elastic materials, stress is proportional to strain, so stress also varies linearly across the depth. The neutral axis passes through the centroid of the cross section, a connection that makes centroid location as consequential here as it is in statics.

The fundamental equation of elastic bending is $\sigma = M \times c / I$, where σ is the bending stress at a distance c from the neutral axis, M is the internal bending moment at that cross section, and I is the moment of inertia about the neutral axis. Maximum bending stress occurs at the extreme fibers, where c is largest.

Section modulus, $S = I / c$, packages this into a single value per cross section: $\sigma_{max} = M / S$. Designers carry tables of S values for standard shapes and pick a section whose S is large enough that σ_{max} stays below the allowable stress.

Bending also produces internal shear. Vertical loads on a beam produce a vertical shear force at every cross section, and this shear generates shear stresses distributed across the section.

For a rectangular cross section, shear stress follows a parabolic distribution, peaking at the neutral axis and dropping to zero at the top and bottom fibers. The maximum shear stress equals 1.5 times the average shear (shear force divided by cross-sectional area). For a wide-flange steel beam, the web carries nearly all the shear and the flanges carry nearly all the bending, so the shear stress is approximated as the shear force divided by the web area. This approximation is accurate enough that codes formally allow it.

The governing differential equation of beam bending connects transverse deflection y to bending moment M through the expression $EI \times d^2y/dx^2 = M(x)$, where E is the modulus of elasticity and I is the moment of inertia. Integrating twice gives the slope and then the deflection along the beam, subject to boundary conditions at the supports. This equation is the foundation of every deflection calculation in structural engineering, whether executed by hand for a simple span, by the moment-area method for a cantilever, or by stiffness-matrix computation for a framed building.

Shear and bending moment diagrams are graphical representations of how shear force and bending moment vary along the length of a beam. They're constructed section by section, sweeping from one end of the beam to the other, adding the effect of each force and distributed load as it's encountered. The shear diagram jumps at every point load, slopes under distributed load (downward if the load is uniformly distributed, curved if it's triangular or parabolic), and closes to zero at the far support. The bending moment diagram is the integral of the shear diagram: it changes slope everywhere the shear is non-zero, reaches local extrema where the shear crosses zero, and closes to zero at the beam ends of a simple span.

Three relationships tie the two diagrams together and the loading diagram to both. The slope of the shear diagram equals the negative of the distributed load intensity. The slope of the moment diagram equals the shear. The change in shear between two points equals the negative of the area under the load diagram between those points. The change in moment between two points equals the area under the shear diagram between them. Once an engineer internalizes these relationships, shear and moment diagrams become almost automatic for any loading configuration.

Locating the maximum moment is the single most important outcome of the diagram exercise, because it's the value against which the beam is designed. For a simple beam with a uniform load w over span L, the maximum moment is $wL^2/8$ at midspan. For a simple beam with a point load P at midspan, the maximum moment is $PL/4$. For a cantilever of length L with a uniform load w, the maximum moment is $wL^2/2$ at the fixed support. These classical formulas reappear throughout design practice, and competent structural engineers have them memorized.

Deflection becomes a serviceability concern when it's large enough to affect function, appearance, or user comfort. Code limits deflection to a fraction of the span length: L/360 for floor members under live load, L/240 for roof members, sometimes tighter for members supporting brittle finishes. A beam that satisfies strength requirements may still fail deflection limits, and the required size is governed by whichever check proves more demanding. Deflection scales with load and span cubed, and inversely with the product of E and I, which is why stiff, deep beams outperform slender ones dramatically when deflection controls.

Shear connections and web stiffeners manage high shear forces in steel beams. Short-span beams with heavy loads, beams in moment frames where shear and moment are both large, and deep beams with thin webs all require detailed shear checks and often need stiffeners to prevent web buckling, web crippling, or local shear failure. These checks operate beyond the simple V/A average shear stress calculation and involve the web's slenderness and support conditions. Similar

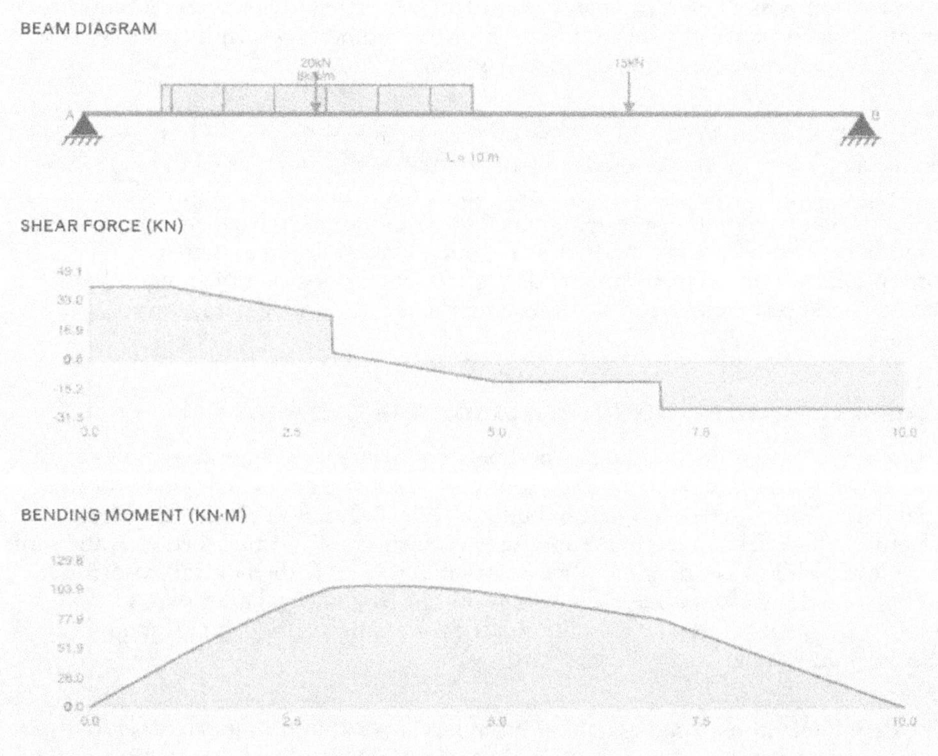

(**Beam Diagram:** A simply supported beam (span ≈ 10 m) carries a **uniformly distributed load (~8 kN/m)** over part of the span, plus two **point loads (≈20 kN and 15 kN)**. Supports at A and B provide reaction forces that balance these loads.

Shear Force Diagram (SFD): Starts positive at the left support, decreases linearly under the distributed load, and shows **sharp drops at point loads** (discontinuities). It crosses toward negative and remains constant where no loads act.

Bending Moment Diagram (BMD): Begins at zero, rises smoothly (parabolic under UDL), peaks near midspan, then decreases. **Slope changes at point loads**, and the moment returns to zero at the right support, consistent with simply supported boundary conditions.)

---

refinements exist for reinforced concrete beams, where shear is resisted by the combined action of the concrete section and stirrups placed at calculated spacings.

Composite beam action occurs when a concrete slab is mechanically connected to a steel beam through shear studs welded to the top flange. The studs force the slab and beam to deflect together, and the slab acts as the top compression flange of the combined section, dramatically increasing both strength and stiffness. A properly designed composite beam can carry 50 to 100 percent more load than a bare steel beam of the same weight, which is why steel-framed buildings with concrete floor slabs almost always use composite construction.

Prestressed concrete beams introduce an additional bending moment opposite to the service load moment by pretensioning or posttensioning high-strength steel tendons embedded in the concrete. The prestress holds the concrete in compression at locations that would otherwise experience tension under service load, allowing the beam to carry much larger loads or span much longer distances than a conventionally reinforced beam could. Prestressed girders span the longest highway bridges, and prestressed one-way slabs underlie many parking garages and commercial buildings.

Trusses, Frames, and Indeterminate Structures

A **truss** is an assembly of straight members connected at their ends to form a geometrically stable arrangement of triangles. Each member carries load only in axial tension or compression, not in bending. This restriction yields exceptional structural efficiency: an axial member uses every fiber of its cross section at the same stress level, while a bending member wastes material near the neutral axis where stress is low. Trusses span large distances with far less material than solid girders would require, which is why they dominate roof framing, bridge main spans, and transmission towers.

The defining feature of a truss is that members meet at joints modeled as pin connections incapable of transmitting moment. Real truss joints are welded or bolted and do transmit some moment, but the secondary stresses from this fixity are typically small compared to the primary axial forces and are neglected in routine design. Loads applied directly to the joints transmit through the members as pure tension or compression. Loads applied at midspan of a member introduce bending that violates the ideal truss assumption, so truss geometry is usually arranged so all significant loads arrive at joints.

The method of joints solves a truss by writing the two equations of equilibrium ($\Sigma Fx = 0$ and $\Sigma Fy = 0$) at each joint and solving for member forces one joint at a time. The analyst starts at a joint with only two unknown member forces, solves for those, moves to the next joint whose unknown count has now dropped to two, and continues until every member is known. The method is systematic and well-suited to hand calculation on small trusses.

The method of sections cuts the truss along a line passing through at most three members, replaces the cut members with their unknown axial forces, and writes three equations of equilibrium ($\Sigma Fx = 0$, $\Sigma Fy = 0$, $\Sigma M = 0$) for the resulting free body. Solving those three equations gives the three unknown member forces in a single step, without having to work through the intermediate joints. The method of sections is especially efficient when only a few member forces are needed, as in preliminary design or when checking the most heavily loaded members.

Truss stability requires that the geometric arrangement actually be rigid. A square frame of four pinned members is not a truss because it deforms freely into a

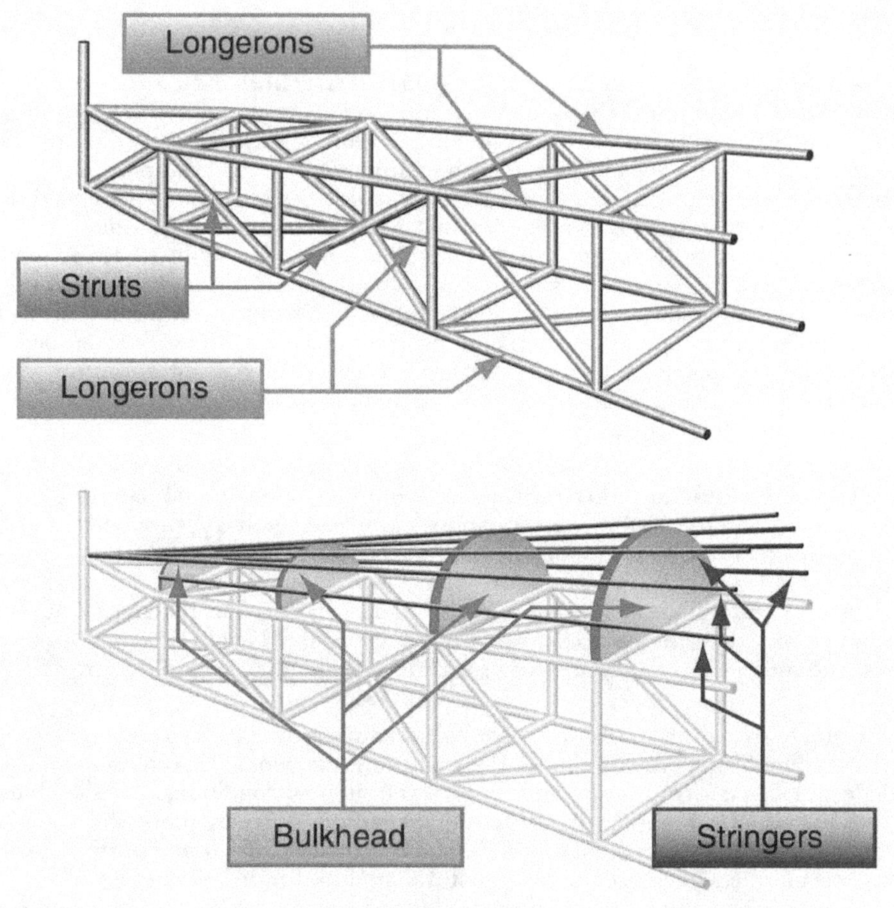

(This image shows a truss-like fuselage where **longerons and struts form triangular load paths**, consistent with the idea that members carry primarily axial tension or compression rather than bending. The **bulkheads and stringers distribute loads to joints**, reinforcing the principle that efficient truss systems route forces through connections to preserve axial behavior and structural efficiency.)

parallelogram under any lateral load. Adding a diagonal member triangulates the square and makes it rigid. The counting rule for planar trusses says that for m members, j joints, and r reaction components, stability and determinacy require $m + r = 2j$. Fewer members than $2j - r$ means the truss is a mechanism: it can move without deforming its members. More members than $2j - r$ means the truss is statically indeterminate.

A **frame** is a planar or three-dimensional assembly of members that carries load through combined axial, shear, and bending action. Unlike trusses, frame members have rigid moment-resisting connections, and the loads on members can act anywhere along their length, not just at joints. Building structures are almost always frames, and the analysis must track axial force, shear, and bending moment at every cross section of every member.

For simple frames, hand methods work. The **portal method** and **cantilever method** are approximate techniques for lateral load analysis of multi-story frames, making simplifying assumptions about the distribution of shear among columns and the location of inflection points within each member. These methods produce reasonable answers for preliminary design and provide the intuition experienced engineers use to sanity-check computer results. Most production work today, however, uses matrix stiffness analysis executed by commercial software. Every joint in a three-dimensional frame has six degrees of freedom (three translations and three rotations), and every member connects two joints with a stiffness matrix describing how end forces and moments relate to end displacements. The software assembles the global stiffness matrix, applies the loads as a force vector, and solves for the displacements. Once displacements are known, member end forces and internal force distributions follow directly.

Statically indeterminate structures have more reaction components or internal forces than the equilibrium equations alone can resolve. A propped cantilever (fixed at one end, supported on a roller at the other) has four reaction components but only three equations of equilibrium, making it indeterminate to the first degree. A continuous beam spanning three supports has four reactions and three equations, also indeterminate to the first degree. Fixed-fixed beams, continuous frames, and most real building structures are indeterminate to multiple degrees.

Indeterminate structures cannot be solved from equilibrium alone. Additional equations come from compatibility of deformations: the geometric requirement that the deflected shape is continuous and respects the support conditions. Classical hand methods include the method of consistent deformations (force method), the slope-deflection method (displacement method), and moment distribution (Hardy Cross's iterative technique, which dominated structural analysis for decades before computers). Each method applies compatibility in its own way, but all produce the same result for a given structure.

Indeterminacy changes how structures respond to load. In a determinate structure, a change in support conditions, an imposed displacement, or a thermal expansion simply translates into different reactions or a different geometry, without internal forces. In an indeterminate structure, the same changes generate internal forces because the structure is resisting the imposed deformation. A continuous beam that experiences support settlement develops new bending moments throughout its length. A rigid frame that experiences a thermal expansion develops axial forces and moments as the connections resist the expansion. This sensitivity to non-load effects is the defining behavioral difference between determinate and indeterminate structures, and it demands care during analysis.

Indeterminacy brings benefits too. An indeterminate structure has multiple load paths: if one element yields or partially fails, load redistributes through alternative paths to supports. This redundancy underlies structural robustness and resistance to progressive collapse. A simply supported beam failing in midspan has no alternate path and collapses. A continuous beam yielding at a support can form a plastic hinge that redistributes the moment to midspan, and the beam continues to carry load at a higher deflection until it forms additional plastic hinges sufficient to make it a mechanism.

This plastic redistribution is the basis of **plastic analysis**, a method that computes the collapse load of a ductile structure rather than its elastic yield load. Plastic analysis takes advantage of the ductility that steel and properly detailed reinforced concrete provide, and it gives collapse loads substantially higher than elastic yield loads for indeterminate structures. Design codes permit plastic design for compact, ductile members and require capacity design procedures that force yielding to occur in designated ductile zones while other elements remain elastic.

Three-dimensional analysis is now routine. A building's lateral load resisting system, shear walls, moment frames, braced frames, responds to wind and seismic loads as a three-dimensional system with coupled horizontal, vertical, and torsional behavior. Proper modeling captures floor diaphragm stiffness, the interaction among multiple lateral systems in the same building, accidental torsion, and vertical irregularities that concentrate demand at particular stories. Software handles the computation, but the engineer is responsible for building a model that reflects the actual structure faithfully and for checking the results against independent sanity checks.

Limit State Design and Safety Factors

Modern structural codes are built around the concept of **limit states**, conditions beyond which the structure no longer satisfies its performance requirements. Two families of limit states govern design: strength (also called ultimate) and serviceability.

Strength limit states address the ultimate capacity of the structure: yielding, buckling, fracture, overturning, sliding, and collapse. These are conditions in which the structure fails to carry its intended loads and potentially endangers occupants. Strength limit states are addressed with load combinations that factor loads up by amounts reflecting the uncertainty in each load type and with capacity reduction factors (or material resistance factors) that scale the computed capacity down by amounts reflecting uncertainty in material properties, workmanship, and analytical models.

Serviceability limit states address conditions under which the structure remains safe but no longer functions as intended: excessive deflection, uncomfortable vibration, cosmetic cracking, permanent damage to non-structural components, or intolerable ponding. Serviceability limit states are checked under unfactored service loads rather than factored ultimate loads, because the concern is not about collapse but about everyday performance.

The American system of **Load and Resistance Factor Design** (LRFD), now standard for steel, reinforced concrete, and most new structural design in the United States, explicitly separates the treatment of loads and resistances. The design requirement is that the factored load demand be less than or equal to the factored nominal resistance: $\Sigma\gamma_i Q_i \leq \phi Rn$, where γ_i is a load factor for each load type Q_i, ϕ is a resistance factor, and Rn is the nominal resistance computed from material specifications and member geometry. Load factors typically range from 0.9 to 1.6 depending on the load and combination, reflecting statistical confidence about each load type's variability. Resistance factors range from 0.65 to 0.95 depending on the failure mode, reflecting the consequences of that failure and confidence in the analytical model.

The older system, **Allowable Stress Design** (ASD), compares the actual service-level stress to an allowable stress derived from the material's yield or ultimate stress divided by a single safety factor. ASD is simpler but less transparent about the sources of uncertainty it addresses, and it can't distinguish between loads with very different variability. ASD persists in some specialties (wood, masonry) and remains available as an option in steel design, but LRFD has displaced it as the primary framework.

Eurocode uses a parallel system of partial factors applied separately to loads and materials, arriving at a similar philosophical framework but with its own conventions. Japanese, Canadian, and Australian codes all operate on limit state concepts as well. The global direction is toward more transparent handling of uncertainty, including probabilistic analyses that explicitly compute failure probability rather than relying on deterministic factors to cover it.

Safety factors exist for reasons beyond randomness in loads and material strengths. They cover analytical approximations (a real structure doesn't behave exactly like the model), construction tolerances (fabricated members aren't perfectly straight or perfectly proportioned), workmanship variability, future changes of use that were not anticipated, deterioration over the service life, and consequences of failure that differ among occupancy categories. A hospital carries a higher importance factor than a warehouse because its failure would produce more severe consequences. A structure protecting a nuclear reactor is designed to standards far exceeding routine commercial work because its failure could be catastrophic.

Modern design codes also recognize that not all failures are equal. A ductile failure mode, in which the member yields gradually and provides visible warning, is preferable to a brittle failure mode, in which the member ruptures suddenly without warning. Codes penalize brittle failures with lower resistance factors or higher safety factors, and they specify detailing requirements that force failure into ductile modes. Capacity-designed seismic systems force yielding into selected energy-absorbing elements and protect brittle elements like connections and columns with explicit overstrength factors. Compact-section requirements for steel beams prevent local buckling before full plastic moment capacity is reached. Minimum reinforcement ratios in concrete beams prevent sudden tensile fracture of the steel at first cracking.

The design engineer operates within this framework, selecting load combinations, choosing materials, sizing members, detailing connections, and checking limit states. The deliverables are drawings that convey the structural intent to the builder, calculations that document the analytical basis of the design, specifications that control material quality and construction procedures, and submittal reviews during construction that verify the built reality matches the designed intent. The output of all this work is a structure that performs as intended for its service life, across the many combinations of conditions it will encounter, with margins that acknowledge everything we cannot predict.

> **To what extent are safety factors a substitute for uncertainty, and where do they fall short?**

Chapter 5: Geotechnical Engineering and Soil Mechanics

Soil is the material every structure ultimately rests on, and it behaves unlike any of the manufactured materials covered earlier. Concrete and steel arrive on site with specifications, certificates of analysis, and predictable properties. Soil was there first. It varies from site to site, from foot to foot within a site, from wet season to dry season, and from undisturbed to compacted condition. Geotechnical engineering is the discipline that studies this variable material and predicts how it will perform when loaded by foundations, cut by excavations, retained by walls, or displaced by tunneling. Without reliable geotechnical analysis, every structure above it would be a guess.

Three realities shape the discipline. Soil is heterogeneous: samples taken ten feet apart often differ in composition, density, and water content enough to matter. Soil is time-dependent: loads applied today produce settlements that may take years to develop. And soil behavior depends on the presence and movement of water, which transforms a material from stable to catastrophic in moments. These three facts drive the methods, the investigations, and the safety factors that distinguish geotechnical engineering from structural work.

Soil Classification and Index Properties

Before an engineer can predict how soil will behave, she has to know what kind of soil it is. **Soil classification** is the language geotechnical engineers use to describe materials quickly and consistently.

Two classification systems dominate: the Unified Soil Classification System (USCS) used in building and general construction, and the AASHTO classification used in transportation work. Both divide soils by grain size and by behavior, then give each category a letter designation that appears on every boring log and foundation drawing.

Soils divide first by grain size. **Coarse-grained soils** contain more than half their material coarser than the No. 200 sieve (0.075 mm). Gravels run from 4.75 mm up to 75 mm. Sands span 0.075 mm to 4.75 mm. **Fine-grained soils** have more than half their material finer than the No. 200 sieve and divide further into silts and clays based on plasticity rather than size alone. Silts have particle sizes between 0.002 and 0.075 mm but don't exhibit plasticity when mixed with water. Clays, often smaller than 0.002 mm, develop plasticity because their plate-shaped particles adsorb water onto their surfaces.

Grain-size distribution characterizes coarse soils and is determined by sieve analysis, passing a sample through a stack of progressively finer sieves and weighing the material retained on each. The resulting distribution curve reveals whether the soil is **well-graded**, containing a range of sizes that pack densely, or **poorly graded**, containing only a narrow range. Well-graded granular fill compacts to higher densities and supports heavier loads than poorly graded fill of the same parent material.

Fine-grained soils cannot be separated by sieve, and their behavior has little to do with grain size anyway. For clays and silts, the **Atterberg limits** define the engineering character of the soil.

The **liquid limit** (LL) is the water content at which the soil transitions from a plastic to a liquid state. The **plastic limit** (PL) is the water content below which the soil ceases to behave plastically and crumbles when rolled into threads. The difference, LL minus PL, is the **plasticity index** (PI), and it quantifies how much water range separates solid from liquid behavior. High-plasticity clays, with PI values above 40, shrink, swell, and cause foundation distress on a scale no other soil matches. Low-plasticity silts, with PI below 7, behave more like structural fill but are vulnerable to erosion and frost heave.

Other index properties tie classification to field performance. **Unit weight** is the weight per unit volume and determines how much load a soil mass imposes on whatever lies beneath it. Typical unit weights range from 100 pcf for dry loose sands to 140 pcf for compacted well-graded granular fill. **Water content** is the mass of water divided by the mass of dry solids, expressed as a percentage. **Void ratio** is the volume of voids divided by the volume of solids, and it tracks how much air or water can occupy the pore space between grains. **Relative density** expresses the actual void ratio against the maximum and minimum possible values for a given soil, giving a measure of how densely the granular particles are packed.

Specific gravity of soil solids typically falls between 2.60 and 2.80 for mineral soils, higher for iron-rich materials and lower for organic contents. **Organic content** itself matters because organic soils compress substantially under load, decompose over time, and perform poorly as foundation material. Peats and organic silts are identified in the classification as OL, OH, or Pt and trigger special design considerations.

Field identification by experienced engineers uses simple tests: rolling a thread of moist soil between the palms to estimate plasticity, shaking a small ball of moist soil to see if water migrates to the surface, rubbing a pinch between the fingers to feel for grittiness versus smoothness. These tests don't replace laboratory analysis but do allow quick judgments in the field during a site visit or subsurface investigation.

The USCS designation for each soil consists of two letters. GW is well-graded gravel. SP is poorly graded sand. CL is low-plasticity clay. MH is high-plasticity silt. An engineer reading a boring log sees a column of these designations and instantly understands the general character of the subsurface, where the water table sits, where strength will be high or low, and where settlement or expansion might become problems.

Effective Stress, Pore Pressure, and Consolidation

The most consequential concept in soil mechanics is **effective stress**. Soil carries load through the contact forces between its solid grains, not through the water in its pore spaces. When an engineer computes the stress acting at a point within a soil mass from the weight of overlying material and applied loads, that total stress divides between the effective stress carried by the grain skeleton and the pore water pressure carried by water in the voids. Karl Terzaghi's effective stress principle, articulated in the 1920s and foundational to modern soil mechanics, states the relationship as $\sigma' = \sigma - u$, where σ' is effective stress, σ is total stress, and u is pore water pressure.

This simple equation has enormous consequences. Consider a saturated clay at 20 feet depth under a new building. The total stress at that point might be 3,000 psf from the weight of overlying soil plus the newly applied load. If the pore water pressure equals 1,200 psf, the effective stress is 1,800 psf. Only the effective stress compresses the soil skeleton and controls its strength. If the water pressure rises to 2,000 psf without any change in the applied load, effective stress drops to 1,000 psf, the soil loses strength, and slopes may fail, excavations may collapse, and foundations may move.

Pore water pressure changes for multiple reasons. The static water table establishes a baseline pressure that rises with depth below the water table. Seasonal fluctuations, rainfall infiltration, adjacent construction dewatering, and tidal effects all shift the baseline. Rapid loading, such as placing a fill on saturated clay, transfers load immediately to the pore water because the soil grains can't rearrange fast enough to accept it. This excess pore pressure dissipates over time as water flows out of the soil into adjacent more permeable layers, and the load gradually transfers from the water to the grains.

That gradual transfer is **consolidation**. When a load is placed on a saturated soil, the initial response puts all the added stress into the pore water, which then bleeds out through whatever drainage paths are available. As water leaves, the grains settle closer together, voids decrease, and the effective stress grows until it carries the full applied load. The settlement that accompanies this process has three components: immediate (or elastic) settlement, primary consolidation settlement, and secondary compression.

Immediate settlement occurs at the moment the load is applied and represents distortion of the soil skeleton before any drainage has happened. For granular soils, immediate settlement is essentially all the settlement, because water drains through gravel and sand almost instantly. For saturated clays, immediate settlement is a small portion of the total.

Primary consolidation is the time-dependent settlement associated with pore water dissipation in fine-grained soils. It dominates in saturated clays. The time to reach a given percentage of consolidation depends on the soil's **coefficient of consolidation** (cv), the drainage path length squared, and a dimensionless time factor. A 10-foot-thick clay layer draining both upward and downward at moderate cv might reach 90 percent consolidation in a year. The same layer with only one-sided drainage reaches the same point in four times as long, because the drainage path doubles and the time scales with the square.

Secondary compression is the additional settlement that continues after primary consolidation is essentially complete. It results from slow rearrangement of the soil fabric under sustained effective stress and generally follows a logarithmic rate of change with time. Secondary compression is negligible for most granular soils but can add a measurable fraction of the primary settlement for highly organic soils and some soft clays over decades.

Consolidation analysis rests on **Terzaghi's one-dimensional consolidation theory**. The theory treats the clay as a porous medium in which water flows according to Darcy's law driven by hydraulic gradients from excess pore pressure.

The governing partial differential equation, $\partial u/\partial t = cv \times \partial^2 u/\partial z^2$, describes how excess pore pressure dissipates with time and depth. Solving it for various boundary conditions gives the time-settlement relationship engineers use to predict how a structure will settle and how long the settlement will take.

The settlement magnitude itself comes from a compressibility relationship determined in the laboratory. A **consolidation test** (or oedometer test) loads a disc-shaped soil sample through a series of stress increments while measuring vertical strain. The resulting stress-strain curve, usually plotted on semi-logarithmic axes, reveals two parameters: the **compression index** (Cc), which governs settlement of **normally consolidated** soil, and the **recompression index** (Cr), which governs settlement of **overconsolidated** soil below its historical maximum stress. The **preconsolidation pressure** (σp'), the maximum vertical effective stress the soil has experienced in the past, marks the transition between the two. Knowing whether current and anticipated stresses are below or above σp' controls whether Cc or Cr applies and often changes the predicted settlement by an order of magnitude.

Quick check against the intuition: placing a 20-foot fill on a 30-foot-thick soft clay layer can produce a foot or more of long-term settlement, with most of it occurring over several years. Engineers manage such settlement through preloading, wick drains that shorten drainage paths, staged construction that allows strength to develop before additional load is placed, or simply avoiding the problem by using a different foundation concept.

Shear Strength and Slope Stability Analysis

Soil fails along surfaces where the shear stress exceeds the available shear strength. Every slope stability problem, every bearing capacity calculation, every retaining wall design depends on predicting shear strength correctly.

Two parameters control it: **cohesion** (c) and the **angle of internal friction** (ϕ). Coulomb's failure criterion, the foundation of every shear strength calculation, states that the shear strength at a point equals c plus the normal effective stress times the tangent of ϕ: $\tau f = c' + \sigma' \times \tan(\phi')$.

Granular soils (sands and gravels) are called **cohesionless** because their strength comes almost entirely from friction between grains. Their cohesion c' is essentially zero, and their friction angle ranges from about 28 degrees for loose clean sand to 45 degrees for dense angular well-graded gravel. Loading a cohesionless soil with no normal stress produces no shear strength at all, which is why sand piles into conical mounds with slopes matching the friction angle rather than standing as vertical cuts.

Clays behave differently. Fine-grained soils develop cohesion from the molecular forces between their plate-shaped particles and from the suction of water bound to particle surfaces. A saturated clay under undrained conditions, loaded too quickly for pore water to escape, exhibits an **undrained shear strength** (su) that's independent of the applied normal stress. This behavior is often called the $\phi = 0$ case because a plot of shear strength versus normal stress appears as a horizontal line. A stiff clay may have su around 2,000 psf; a soft clay, around 250 psf; a very soft marine clay, less than 100 psf.

The difference between **drained** and **undrained** conditions is central to clay behavior. Drained conditions prevail when the loading is slow enough (or the soil permeable enough) that pore water pressure stays at equilibrium throughout loading. Undrained conditions apply when loading is fast enough that pore pressure develops and can't dissipate. Most short-term problems in saturated clays (end of construction, rapid drawdown of a reservoir, earthquake loading) use undrained strengths. Long-term stability problems in the same soils use drained strengths, because given enough time the excess pore pressures equilibrate.

Shear strength is measured in the laboratory using several standard tests. The **direct shear test** forces a sample to fail along a predefined horizontal plane while applying a normal load, and the resulting shear-versus-normal stress plot gives c' and ϕ' directly. The **triaxial test** subjects a cylindrical sample to a controlled confining pressure, then applies an axial load until the sample fails. Triaxial tests can be run consolidated-drained (CD), consolidated-undrained (CU), or unconsolidated-undrained (UU), each replicating different field conditions. Unconfined compression tests on intact clay samples provide a quick measure of undrained strength for preliminary analysis.

Slope stability analysis predicts whether a natural slope, an engineered embankment, or the side of an excavation will remain in place. Classical methods assume a failure surface within the soil mass, compute the shear stress acting along that surface from the weight of the sliding mass and the applied loads, compute the shear strength available along the same surface from the soil properties and the normal stresses, and compare the two in a **factor of safety**:

FS = available resisting moment or force divided by driving moment or force.

Factors of safety above 1.5 are typical for permanent slopes under long-term drained conditions, with higher values for critical structures or conditions with high uncertainty.

The failure surface is rarely known in advance, so analysis must search for the critical surface, the one with the lowest factor of safety. For circular slip surfaces in homogeneous soils, **Bishop's simplified method** and **Spencer's method** divide the sliding mass into vertical slices, compute forces on each slice, and solve iteratively.

Non-circular surfaces apply in layered or complex soils, and modern software routinely searches over thousands of potential surfaces to find the critical one.

Infinite slope analysis applies to long, shallow slopes where failure is dominated by sliding parallel to the slope surface rather than rotational failure within the soil mass. Road cuts in weathered bedrock, long embankments, and surface soils on hillsides often fail this way, and the analysis reduces to checking the ratio of shear strength to shear stress along a plane parallel to the slope at the expected depth of failure.

Slope stability problems have characteristic failure mechanisms that depend on soil type and geometry. Rotational failures, with nearly circular slip surfaces, occur most often in homogeneous cohesive soils. Translational failures, with nearly planar slip surfaces, predominate in stratified soils with a weak layer at depth. Flow failures, where the soil mass behaves like a fluid after failure, happen in very soft saturated clays and in liquefied sands under earthquake loading. Retrogressive failures start at the toe of a slope and progress backward into the slope through successive slips. Each mechanism demands the right analytical approach and the right countermeasures.

Liquefaction deserves separate mention. Loose saturated sands and silty sands under cyclic loading from an earthquake can lose their grain-to-grain contact as pore pressure builds toward the confining stress. When effective stress drops to near zero, the soil temporarily loses almost all strength and behaves as a heavy fluid. Buildings sink, basement walls float, tanks tilt, and bridges whose foundations rested on liquefiable soil collapse. Engineers evaluate liquefaction potential from the soil's relative density, depth, and the anticipated ground motion, and they mitigate through densification, drainage to relieve pore pressure, grouting, or supporting structures on deep foundations passing through the liquefiable layers.

Shallow and Deep Foundation Systems

Foundations transfer structural loads from the superstructure into the soil below. The choice of foundation type depends on the soil profile, the loads involved, the allowable settlement, the depth to adequate bearing, and cost. Foundations divide broadly into shallow and deep systems.

Shallow foundations distribute structural loads over the soil immediately beneath the building, typically within a depth of a few feet to a few dozen feet. The most common types are spread footings, strip footings, mat foundations, and combined footings. A **spread footing** is an individual block of concrete, usually square or rectangular, placed under a single column. A **strip footing** is a continuous ribbon placed under a load-bearing wall. A **mat foundation** (or raft) is a large slab covering the entire building footprint, used when individual footings would occupy so much of the area that a continuous slab becomes more efficient, or when the soil is so weak that only a very wide bearing area provides adequate support.

The design of a shallow foundation addresses three questions. First, what is the **bearing capacity** of the soil: the maximum pressure that can be applied without causing shear failure of the soil beneath the footing? Second, what is the **settlement** that will result from the applied pressure, and is it acceptable? Third, is the foundation itself, the concrete and reinforcement, adequate to transmit the loads without failing structurally?

Bearing capacity theory originated with Karl Terzaghi's equation for continuous strip footings on cohesive or cohesionless soil: $q_{ult} = c'N_c + q'N_q + 0.5\gamma BN_\gamma$, where c' is cohesion, q' is the overburden pressure at the footing base, γ is the unit weight below the base, B is the footing width, and N_c, N_q, and N_γ are bearing capacity factors that depend on the friction angle. Meyerhof, Hansen, and Vesic extended Terzaghi's original equation with shape factors, depth factors, inclination factors, and ground slope factors to handle the many real-world variations. Allowable bearing capacity divides the ultimate value by a safety factor of 2.5 to 3.0.

Settlement analysis for shallow foundations on granular soils uses simplified elastic methods or correlations with standard penetration test (SPT) blow counts. For foundations on cohesive soils, the consolidation theory from earlier in this chapter applies directly: the added pressure from the footing creates stress changes at depth, those stress changes produce consolidation settlement in compressible layers, and the total settlement is the sum of contributions from each layer. Differential settlement between adjacent footings often matters more than absolute settlement, because differential movement cracks walls, tilts structures, and damages utilities in a way that uniform settlement does not.

Deep foundations transfer loads to competent strata below soft or compressible near-surface soils. They come in two primary families: piles and drilled shafts. **Piles** are slender structural members (steel, concrete, or timber) driven into the ground by impact or vibratory hammers. **Drilled shafts** (also called caissons or bored piles) are cast-in-place concrete columns formed by drilling a hole to the required depth and filling it with concrete and reinforcement.

Piles and drilled shafts carry axial load through two mechanisms: **skin friction** along their shaft and **end bearing** at the tip. The total capacity is the sum of the two, each multiplied by the appropriate factor. Friction piles rely predominantly on side friction and are useful where no firm bearing stratum exists within economical depth. End-bearing piles pass through weak soils to a firm layer that carries nearly all the load at the tip. Most real piles draw on both mechanisms in varying proportions.

Pile capacity is estimated by static formulas based on soil properties, by dynamic formulas based on driving resistance, or by full-scale load tests that apply known loads and measure the resulting deflections. Load tests are the most reliable but also the most expensive, and they're typically performed on large projects as a verification of the design assumptions. PDA (Pile Driving Analyzer) testing, performed during installation, uses strain and acceleration measurements to estimate capacity dynamically without a separate static test.

Pile **groups** behave differently from single piles. Piles spaced closely influence each other's stress zones in the soil, and the group capacity is less than the sum of individual pile capacities, though the reduction varies with pile spacing and soil type. A group of piles also causes settlement by imposing a loaded area on deeper soil layers, and the analysis must account for this group settlement in addition to the individual pile behavior.

Micropiles, small-diameter drilled piles grouted under pressure, are used where conventional equipment can't access the site, where existing foundations need underpinning, or where light structures need cost-effective deep support. They rely almost entirely on skin friction and gain their capacity from the bond between grout and surrounding soil. **Helical piles** (screw piles) have helical plates welded to a central shaft and are screwed into the ground by torque, with capacity related to the torque required for installation.

Foundation selection emerges from balancing all these factors. A small single-family home on good soil takes spread footings at the bottom of the frost line. A mid-rise building on deep soft clay takes drilled shafts or driven piles extending to firm bearing below. A highway bridge pier in a riverbed takes large diameter drilled shafts embedded in bedrock. An LNG storage tank on soft soil takes a mat foundation combined with stone columns or deep soil mixing to improve the underlying clay. Every project finds its own answer.

Earth Retention Structures and Retaining Walls

When a site cannot provide a gentle slope between elevations, a retaining structure holds back the higher ground and keeps soil from sliding down onto the lower level. Retaining walls appear in highway cuts and fills, basement walls of buildings, wharves and seawalls, bridge abutments, and countless other applications. They all resist lateral earth pressure, but they do so in distinct structural ways.

Lateral earth pressure is the horizontal force the soil mass exerts on whatever restrains it. Its magnitude depends on the density and strength of the soil, the height of the retained material, the inclination of the wall and ground surface, and critically, on whether the wall can yield or is held rigid. Three earth pressure states apply. The **active state** develops when the wall yields slightly outward, allowing the soil behind it to reach a failure condition that minimizes the lateral force: this is the lowest pressure and applies to free-standing retaining walls that can deflect. The **at-rest state** describes an unyielding wall, such as a basement wall restrained by adjacent floors, and produces higher pressures than the active state. The **passive state** develops when the wall is pushed into the soil, mobilizing the soil's maximum resistance, and produces pressures several times higher than either active or at-rest.

Coulomb and Rankine developed the classical theories for active and passive earth pressure. Rankine's theory assumes a vertical wall with horizontal backfill and gives clean formulas: the active earth pressure coefficient Ka equals $\tan^2(45° - \phi/2)$, while the passive coefficient Kp equals $\tan^2(45° + \phi/2)$. For a soil with $\phi = 30$ degrees, Ka = 0.33 and Kp = 3.0, a factor-of-nine difference in the two states. Coulomb's theory accepts inclined walls, sloped backfills, and wall friction, producing more general formulas at the cost of complexity. Both theories inform the code-based methods engineers use today.

Retaining wall types align with the earth pressure the wall must resist and the height involved. **Gravity walls**, built from massive concrete or stone, resist overturning and sliding through their own weight. They're economical for heights up to 10 to 15 feet. **Cantilever retaining walls** are reinforced concrete structures with an L or T shape that uses the weight of the soil on the heel to stabilize against overturning. They work efficiently for heights up to 25 feet or so. **Counterfort walls** add

(This image shows a **stone gravity retaining wall** at the Point Reyes Lighthouse, built into a steep coastal slope to stabilize the terrain. Its large mass resists the lateral forces from the retained soil and rock. This prevents sliding and protects the structures above from erosion and slope failure.)

vertical stiffening ribs on the back of tall cantilever walls, extending the economical range upward.

Mechanically stabilized earth (MSE) walls use layers of tensile reinforcement (galvanized metal strips, welded wire mesh, or geosynthetic straps and grids) embedded in compacted granular backfill. The reinforcement extends from a facing element back into the retained soil, and the composite mass behaves like a gravity wall at a fraction of the cost per foot of height. MSE walls are the workhorse of highway construction and appear on virtually every interchange built in the last 30 years.

Sheet pile walls are interlocking steel or vinyl sections driven into the ground, cantilevering above the excavation line or anchored to external tiebacks or internal struts. They're common in waterfront construction, cofferdams, and temporary excavation support. **Soldier pile and lagging** walls combine widely spaced vertical steel soldier piles with horizontal wood lagging between them, making them inexpensive for temporary earth retention during building construction.

Soil nail walls are formed by drilling and grouting steel bars into an existing soil face and then covering the face with shotcrete reinforced with wire mesh. The nails extend beyond the failure surface and tie the soil back to the stable zone. They can be installed from the top down as an excavation proceeds, without the need to construct a wall at the bottom first. They work well in stiff clays and dense granular soils where the soil can hold an open cut between nail installations.

Design of every retaining structure addresses multiple failure modes. Overturning about the toe must be resisted by the righting moment from the wall and any soil weight above it. Sliding along the base must be resisted by friction and any embedment into passive soil ahead of the toe. Bearing pressure at the base must stay within the soil's allowable capacity. Global (deep-seated) slope stability, with a failure surface passing behind and beneath the wall, must show an acceptable factor of safety. And the internal structural components of the wall must carry their own design loads with appropriate safety margins. A wall that passes four checks but fails one still fails.

Ground Improvement and Soil Stabilization Techniques

Sometimes the existing soil simply won't do. Rather than designing around it with costly deep foundations or oversized footings, engineers can modify the soil itself. **Ground improvement** encompasses techniques that densify, drain, reinforce, or strengthen soil in place, often at a fraction of the cost of deep foundations.

Compaction is the simplest and most widely applied form of ground improvement. When soil is placed as structural fill, it's compacted in thin lifts by rollers (smooth drum, sheepsfoot, or pneumatic) to increase its density. Proctor tests in the laboratory establish the maximum dry density achievable for the soil at various water contents, and field compaction specifications require a minimum percentage of that maximum, typically 95 percent for structural fill. Vibratory rollers work well for granular soils. Sheepsfoot rollers work better for cohesive soils, kneading the clay and breaking up any loose lumps.

Dynamic compaction densifies existing loose granular soil in place by repeatedly dropping a heavy weight (5 to 30 tons) from a height of 40 to 100 feet. The impacts create craters that are backfilled and recompacted, and the energy propagates through the soil mass to depths of 20 to 30 feet. It's a relatively crude but economical technique for large sites with loose sands or miscellaneous fill.

Vibroflotation densifies saturated granular soils using a vibrating probe that's jetted or vibrated into the ground and then slowly withdrawn. The vibration causes the sand grains to rearrange into a denser configuration. **Vibro-replacement** variants add granular backfill during extraction, forming stone columns that reinforce softer soils and provide drainage paths.

Preloading and surcharging accelerate consolidation settlement by applying a temporary load larger than the final design load. The additional surcharge forces more of the long-term settlement to occur during the preloading period, and when the surcharge is removed before permanent construction begins, the remaining settlement is small. Preloading is often combined with **prefabricated vertical drains** (wick drains), which shorten drainage paths in thick clay layers and reduce the time required for consolidation from years to months.

Deep soil mixing injects cement grout into the soil through rotating augers, blending it with the native material to form columns of soil-cement. The resulting composite is stronger and stiffer than the native soil and can support foundations, stabilize slopes, or provide cutoff walls against seepage.

Jet grouting uses high-pressure jets of cement slurry to erode and mix the soil, creating columns or panels of stabilized material.

Chemical stabilization with lime or cement treats fine-grained soils by chemically altering their plasticity and strength. Lime added to high-plasticity clay reduces the plasticity index, makes the soil workable, and increases long-term strength through pozzolanic reactions. Cement stabilization works on a wider range of soils and produces higher strengths at the cost of more material. Chemical stabilization is standard practice for subgrade preparation in highway construction, where the in-place soil would otherwise perform poorly under pavement loads.

Geosynthetics have transformed ground improvement over the past four decades. Geotextiles (woven or nonwoven polymer fabrics) provide separation between dissimilar soil layers, filtration in drainage systems, and reinforcement under loaded areas. Geogrids (open mesh polymer grids with high tensile strength) reinforce embankments, slopes, and MSE walls, interlocking with granular soil to distribute loads over larger areas. Geomembranes (impermeable polymer sheets) contain fluids in landfills and ponds. Geocomposites combine two or more functions in a single product.

Selection among these techniques depends on the soil type, the depth of improvement required, the time available, the loads to be supported, and environmental constraints. A highway embankment over soft marine clay might combine preloading, wick drains, and stone columns along its length. A building foundation on loose sand might use vibroflotation. A contaminated site might use in-situ soil mixing with cement and additives that both strengthen the soil and immobilize contaminants. The catalog of techniques continues to grow, and competent geotechnical practice today assumes familiarity with all of them, because the right solution often combines two or three in a coordinated program matched to the site's specific demands.

Given that soil conditions are never perfectly known, how do engineers justify confidence in foundation designs?

Chapter 6: Transportation Engineering and Highway Systems

Roads carry the everyday movement that modern economies depend on: commuters driving to work, freight trucks moving goods across continents, emergency vehicles reaching accidents, families traveling to see relatives. A highway is simultaneously a piece of pavement, a geometric alignment through the landscape, a system of signs and signals, and an operational network with predictable traffic characteristics. Transportation engineering addresses all four dimensions together, because a beautifully engineered pavement on a badly aligned road, or a perfectly designed curve with inadequate capacity, produces poor outcomes for the people using it.

The discipline has evolved from its origins in paving technology to a data-rich practice that incorporates traffic simulation, human factors research, environmental analysis, and increasingly, the behavior of connected and autonomous vehicles.

What hasn't changed is the fundamental triangle at the center of every highway project: the pavement that carries the wheel loads, the geometry that guides the vehicle safely around curves and over hills, and the traffic operation that determines how many vehicles can use the facility at what speeds. This chapter covers each of the three.

Pavement Types, Subgrade Design, and Flexible Versus Rigid Systems

Every pavement is a layered structure that distributes concentrated wheel loads over a wide enough area of the underlying soil that the soil doesn't fail or deform excessively. The surface layer resists the wheel directly, provides skid resistance, and sheds water. The intermediate layers reduce stresses with depth and protect the subgrade from moisture and freeze-thaw cycles. The **subgrade** is the native or prepared soil on which the pavement sits. It's the foundation of the whole system, and its strength and behavior govern what the pavement above it must be designed to carry.

Pavement design begins with evaluating the subgrade. The **California Bearing Ratio** (CBR) test is the traditional measure, comparing the resistance of a compacted soil sample to penetration by a standard plunger against the resistance of a reference crushed stone. CBR values range from 2 to 3 for soft clay subgrades unsuitable for direct paving to 30 to 80 for granular materials that perform as subbase or base course. Modern design has shifted toward the **resilient modulus** (Mr), a measure of the subgrade's recoverable deformation under cyclic loading that better represents the repeated stresses pavement actually sees under traffic. Correlations between CBR and Mr appear in design guides, though direct Mr testing is preferred for major projects.

Soil types behave predictably under pavement. Clean granular soils compact well, drain well, and provide good subgrade support. Plastic clays swell when wet, shrink when dry, and transmit those movements upward through the pavement as rutting, cracking, and differential heave. Frost-susceptible silty soils draw water upward to freezing fronts during cold weather, forming ice lenses that heave the pavement and leave voids when they thaw in the spring. Each of these behaviors demands specific design responses: removing and replacing unsuitable soils, stabilizing with lime or cement as described in the previous chapter, or adjusting the pavement structure to accommodate what the subgrade will do.

Two fundamentally different pavement philosophies have developed over the past century. **Flexible pavements** use hot-mix asphalt concrete as the surface layer over a granular base and subbase. They flex under wheel loads, distributing stress through the granular layers into the subgrade by a mechanism resembling a spread footing. When loaded, they deform elastically and recover when the load passes. Over time, cumulative strain gradually consumes the pavement's service life through rutting, fatigue cracking, and surface raveling.

Rigid pavements use Portland cement concrete as the surface layer, with the concrete slab acting as a beam that bridges over areas of weak subgrade support and distributes load through its own bending stiffness rather than through underlying layers. The slab bends under load, but it does so at stresses far below failure, and wheel load stresses transfer mostly through the slab itself to distant supports rather than spreading downward as concentrated pressure on the subgrade.

Each system suits different conditions. Flexible pavements cost less initially, adapt easily to minor settlement, and can be resurfaced cheaply by milling off the top and placing a new overlay. They dominate on state highways with moderate truck traffic, on residential streets, and on pavements where asphalt's lower cost pays off over a 15 to 25 year design life. Rigid pavements last longer (40 years or more is common), handle heavy truck traffic better, resist rutting in hot climates, and hold up to spilled fuel and chemicals that degrade asphalt. They dominate on interstate highways through freezing and thawing northern climates, at truck terminals, and at airport runways and aprons where aircraft tire pressures exceed anything a highway carries.

An **asphalt concrete** pavement section typically includes a wearing course of fine-graded asphalt on top, a binder course of coarser-graded asphalt below, a granular base of crushed stone, and often a subbase of lower-quality granular material on the subgrade. Total thickness ranges from 8 to 12 inches for residential streets to over 24 inches for interstate highways carrying heavy truck traffic. The asphalt mix itself balances the competing demands of rut resistance (achieved with stiff asphalt binders and angular aggregates), fatigue resistance (achieved with softer binders and flexible mixes), low-temperature cracking resistance (also favoring softer binders), and moisture resistance (achieved through careful aggregate selection and anti-strip additives).

Superpave, the current U.S. specification framework, matches binder grades to climate through a performance grade designation like PG 64-22, where the first number is the high-temperature rating in degrees Celsius and the second is the low-temperature rating.

A **Portland cement concrete** pavement section uses a concrete slab typically 8 to 14 inches thick, placed on a prepared base course and subgrade. **Jointed plain concrete pavement** (JPCP) is the most common variant in highway use: plain unreinforced concrete with transverse contraction joints spaced every 12 to 20 feet to control random cracking. **Jointed reinforced concrete pavement** (JRCP) extends the joint spacing to 30 to 100 feet and uses embedded steel mesh to tie the slab across cracks that form between the longer joints. **Continuously reinforced concrete pavement** (CRCP) contains heavy longitudinal reinforcement running continuously through the slab, allowing closely spaced transverse cracks to form in a controlled pattern and eliminating transverse joints entirely. CRCP is the premium system for the highest-volume corridors, delivering the smoothest ride and longest life at the highest initial cost.

Dowel bars and tie bars connect adjacent slabs at joints. **Dowel bars**, smooth round steel bars embedded across transverse joints, transfer wheel loads between slabs without restraining joint movement. **Tie bars**, deformed reinforcing bars at longitudinal joints, hold adjacent lanes together but allow individual slabs to curl and warp naturally under daytime thermal gradients.

Pavement design now uses **mechanistic-empirical** methods rather than the purely empirical AASHTO Road Test methods that dominated for 40 years. The Mechanistic-Empirical Pavement Design Guide (MEPDG, AASHTOWare Pavement ME) models the pavement as a layered elastic system, computes stresses and strains under each axle load in each season, and uses transfer functions to predict distresses (rutting, fatigue cracking, transverse cracking) over the design life. This approach accounts for truck traffic spectra, climatic effects, material variability, and the specific behaviors of different asphalt and concrete mixtures in ways the older methods could not.

Traffic loading for pavement design is quantified in **Equivalent Single Axle Loads** (ESALs). A single pass of a standard 18,000-pound single axle equals one ESAL. Heavier axles cause exponentially more damage than lighter ones: a 20,000-pound axle produces roughly 1.5 ESALs, while a 30,000-pound axle produces nearly 8. This fourth-power relationship (damage proportional to load to the fourth) is why a single 80,000-pound truck does more pavement damage than thousands of passenger cars, and why pavement design centers on projecting truck traffic over the design life rather than total vehicle counts.

Horizontal and Vertical Roadway Alignment Geometry

A road's alignment is its path through three-dimensional space, and highway geometric design produces that alignment by combining a horizontal layout (the path projected onto a flat plane) with a vertical profile (the changes in elevation along the path). Both components must respect physical constraints on vehicle behavior, driver perception, and sight lines.

Design speed anchors every geometric decision. Design speed is the chosen speed that governs the tightest curves, steepest grades, and minimum sight distances allowed on the project. For a rural interstate, design speed is typically 70 or 75 mph. For an urban arterial, 45 to 55 mph. For a residential collector, 30 to 35 mph. Once design speed is set, all the geometric criteria follow from it through code tables and formulas in the AASHTO "Green Book" (A Policy on Geometric Design of Highways and Streets).

Horizontal alignment consists of straight tangents connected by circular curves, usually with **spiral transition curves** at each end of the circular portion to give drivers a gradual transition from tangent to full curvature.

The **minimum radius** of a circular curve depends on design speed and the rate of superelevation: $R_{min} = V^2 / (15 \times (e + f))$, where R is radius in feet, V is design speed in mph, e is superelevation rate, and f is the maximum allowable side friction factor.

For a 70-mph design speed with 8 percent superelevation and a side friction factor of 0.10, the minimum radius is about 1,800 feet. Tighter curves force drivers to rely on friction beyond comfortable limits, particularly in wet conditions, and they increase the risk of run-off-road crashes.

Superelevation is the cross-slope of the roadway on a curve, tilted toward the inside of the curve to counteract the centrifugal force acting on vehicles at design speed. Maximum superelevation rates range from 4 percent in urban areas with frequent intersections and pedestrians to 8 or even 10 percent on rural interstates where winter ice is not an overriding concern. The transition from normal crown (typically 2 percent cross-slope sloping from the centerline toward each shoulder) to full superelevation happens over a runoff length that varies with design speed, with the runoff typically occurring within the spiral curve.

Stopping sight distance (SSD) is the minimum distance a driver needs to see an object in the roadway and come to a complete stop before hitting it. It equals the distance traveled during perception-reaction time (conventionally 2.5 seconds) plus the braking distance at the design speed. At 65 mph, SSD is approximately 645 feet. Every curve, every crest vertical curve, and every overpass must provide at least this sight distance to the driver throughout the length of the facility. When the geometry can't provide SSD, warning signs, reduced speed limits, or obstacle removal becomes necessary.

Passing sight distance (PSD) is a greater distance required on two-lane highways where a driver must see far enough ahead to complete a passing maneuver safely. PSD accounts for the time and distance a passing vehicle takes to accelerate into the opposing lane, complete the pass, and return to its own lane before an oncoming vehicle reaches the same point. At 65 mph, PSD exceeds 2,000 feet. Areas that meet SSD but not PSD are marked with no-passing zone striping.

Vertical alignment consists of straight grades connected by parabolic **vertical curves**. A **crest vertical curve** is concave downward, forming the top of a hill. A **sag vertical curve** is concave upward, forming the bottom of a valley or a highway undercrossing. The design parameter for vertical curves is the **K value**, defined as the length of curve divided by the algebraic difference in grades. K values are tabulated in the Green Book for each design speed, calibrated to provide adequate stopping sight distance on crests and adequate comfort on sags.

For a crest vertical curve, sight distance is limited by the curve geometry itself: beyond a certain distance, the driver's line of sight runs into the roadway surface ahead. The minimum curve length to provide SSD depends on the algebraic difference in grades and the driver eye height (3.5 feet) plus the object height (2 feet for stopping, or taller for other design scenarios). For a sag vertical curve, sight distance in daylight is essentially unlimited, but at night it's limited by headlight coverage, and the curve length depends on headlight angle and mounting height.

Maximum grades balance construction cost against vehicle performance. Mountain highways may exceed 6 percent; interstate highways generally limit grades to 3 to 4 percent in mountainous terrain; urban collectors accept up to 8 percent in difficult terrain. Truck traffic imposes the practical constraint: a heavily loaded truck slows dramatically on steep grades, creating speed differentials with passenger cars that both inconvenience drivers and increase crash risk.

Climbing lanes added to long steep upgrades let slow trucks move out of the through lane, restoring capacity and reducing the speed differential problem.

Intersections and interchanges introduce their own geometric requirements. Intersection sight triangles give drivers at stop-controlled or signalized approaches enough view of cross-street traffic to enter safely. Left-turn lanes at signalized intersections require minimum storage length, taper length, and clear throat width. Interchange ramps follow design speeds matched to the traffic they're expected to handle, with acceleration lanes on entrance ramps and deceleration lanes on exit ramps sized to let drivers safely merge into or diverge from the through lanes.

Driver expectancy is an overarching principle that's not a formula but shapes every decision. Drivers develop expectations about what's coming next based on what they've been seeing, and geometric features that violate those expectations surprise drivers and produce crashes. A sharp curve at the end of a long tangent; a decision point where the major movement goes left instead of straight; a ramp that exits on the left rather than the right. Good geometric design keeps the pattern consistent, introduces changes gradually, and provides advance warning when the expected pattern must break.

Traffic Flow Theory, Capacity, and Level of Service

Building a road that carries the expected traffic requires understanding how traffic actually behaves. Traffic flow theory treats vehicles simultaneously as individual drivers making decisions and as a collective stream with measurable aggregate properties. Both views are necessary for design and operation.

Three fundamental traffic stream variables describe any roadway at any moment. **Flow** (q) is the number of vehicles passing a fixed point per unit time, typically expressed in vehicles per hour (vph). **Speed** (v) is the rate at which vehicles travel, measured as either the time-mean speed (arithmetic mean of instantaneous speeds at a point) or the space-mean speed (total distance traveled divided by total travel time, which is the harmonic mean of spot speeds). **Density** (k) is the number of vehicles per unit length of roadway, typically vehicles per mile. The three are connected by the fundamental relationship $q = v \times k$: flow equals speed times density.

Plotting flow against density for a given facility produces a characteristic curve. At low density (few vehicles on the road), drivers travel at free-flow speed and flow is low because there aren't many vehicles even though each moves fast. As density increases, drivers begin to influence each other, speeds drop, and flow rises because the product of speed and density keeps growing.

Eventually, flow reaches a **maximum** called **capacity**, beyond which further increases in density reduce speeds so much that flow actually declines. This point of maximum flow is the defining performance limit of the facility. Beyond capacity, the facility enters **congested flow**, where density continues to rise but flow drops, producing the stop-and-go conditions familiar to every urban commuter.

The **fundamental diagram** that relates flow, speed, and density is roughly triangular or parabolic depending on which theoretical model is used. Greenshields' model, the simplest, assumes a linear speed-density relationship and produces a parabolic flow-density curve with capacity at half the jam density. More sophisticated models reproduce the observed asymmetry between uncongested and congested branches and the scatter observed in real data.

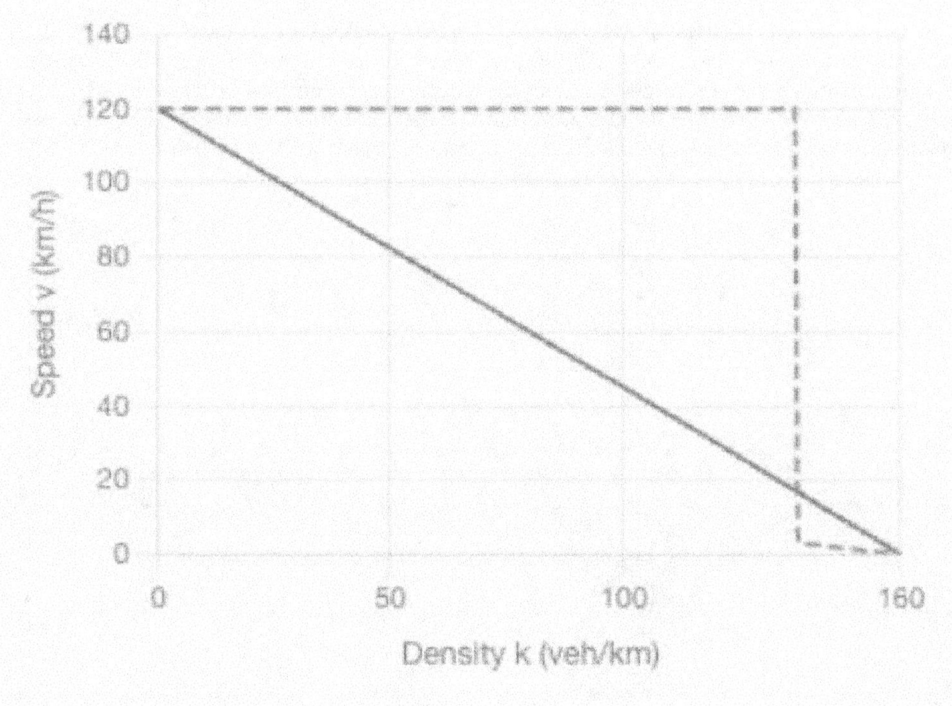

(The speed–density diagram shows two models. The solid line is Greenshields' linear model: speed drops steadily from 120 km/h at zero density to zero at jam density (160 veh/km). The dashed line is the triangular model: speed holds constant at 120 km/h until near jam density, then drops abruptly to zero.)

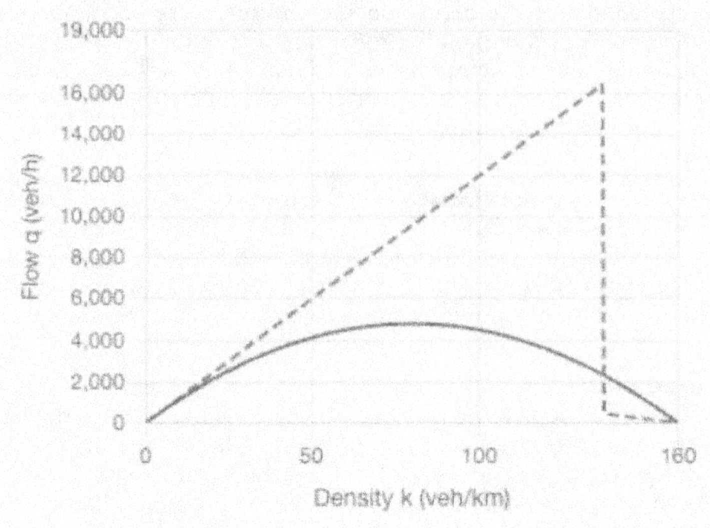

(The flow–density diagram shows the two models diverging sharply. The solid Greenshields curve forms a low, symmetric parabola peaking around 4,800 veh/h at half the jam density (~80 veh/km). The dashed triangular model rises steeply and linearly through the uncongested branch, reaching a much higher capacity of roughly 16,000 veh/h just before jam density, then collapses vertically. This reflects its assumption that free-flow speed is maintained almost until gridlock.)

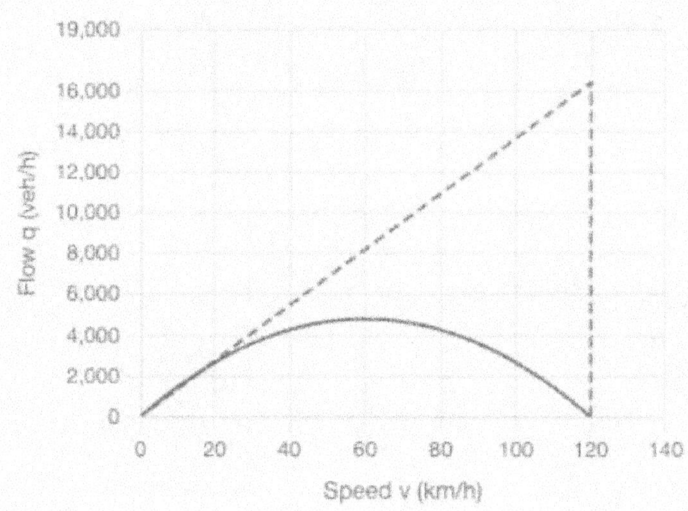

(The flow–speed diagram shows Greenshields (solid) as a symmetric parabola peaking near 4,800 veh/h at around 60 km/h, then declining back to zero at free-flow speed. The triangular model (dashed) rises linearly all the way to 16,000 veh/h at full free-flow speed (120 km/h), then drops vertically (reflecting its abrupt capacity cutoff.))

Highway capacity analysis applies this theory to specific facility types. The Highway Capacity Manual (HCM), currently in its 7th edition, tabulates capacities and defines performance measures for freeways, multilane highways, two-lane highways, signalized intersections, unsignalized intersections, roundabouts, weaving sections, merge and diverge areas, transit facilities, pedestrian facilities, and bicycle facilities. For a basic freeway segment, the capacity ranges from 2,300 to 2,400 passenger cars per hour per lane, with the exact value depending on free-flow speed and lane width. For a signalized intersection approach, capacity depends on the green time available and the saturation flow rate of the lane, both of which involve multiple adjustment factors for heavy vehicles, lane widths, turning movements, and pedestrian interactions.

Level of service (LOS) is the qualitative description of operating conditions that the HCM overlays on the quantitative capacity analysis. Six levels, A through F, describe conditions ranging from free flow (A) to breakdown (F). LOS A means light traffic with drivers free to choose their speeds and maneuvers. LOS B adds mild interaction with other vehicles but still unrestricted speeds. LOS C represents stable flow with maneuvering restricted and speeds approaching the free-flow speed. LOS D is high-density stable flow where most drivers can't move freely in their desired lane. LOS E operates at or near capacity, with speeds significantly reduced and any disruption causing queues. LOS F is breakdown: demand exceeds capacity, speeds and flow both fall, queues form and lengthen until demand subsides.

The specific performance measure defining each level varies by facility. For freeway segments, LOS is based on density (passenger cars per mile per lane). For signalized intersections, it's based on control delay (average seconds of delay per vehicle). For two-lane rural highways, it's based on percent time spent following (the fraction of time drivers travel within 3 seconds of the vehicle ahead). Different facilities require different measures because what defines "quality of service" differs by context.

Most urban streets operate under **signal control**, and the analysis of signalized operations deserves its own attention. A signal cycle divides into green, yellow (typically 3 to 6 seconds), and red phases for each approach. The green time allocated to each approach is typically proportional to the demand on that approach.

Saturation flow rate, the rate at which queued vehicles discharge across the stop bar after the signal turns green, is approximately 1,900 vehicles per hour of effective green per lane under ideal conditions, adjusted downward for heavy vehicles, parking, buses, and turning movements. The **effective green time** is the actual green time plus the yellow time, minus start-up lost time (2 to 3 seconds at the beginning of each phase). The approach capacity equals saturation flow times effective green divided by cycle length.

Webster's formula for optimum cycle length at a signalized intersection gives $C\_opt = (1.5 \times L + 5) / (1 - \Sigma y)$, where L is the total lost time per cycle and Σy is the sum of critical lane group flow ratios. This formula, developed by F. V. Webster in the 1950s and still quoted in textbooks, balances delay between vehicles forced to wait through red phases and vehicles slowed by inefficient short cycles. Modern signal optimization software can refine cycle length, phase splits, and offsets across a network to minimize overall delay or maximize throughput.

Coordination between adjacent signals along an arterial produces the familiar **green wave** where drivers traveling at a target speed encounter a series of green signals. Coordination requires common cycle lengths across all signals in the coordinated group and carefully chosen offsets between them. Simple two-way coordination on moderate-demand arterials is straightforward.

Complex grid coordination with high demands in multiple directions forces compromises, since optimizing for one direction often degrades operations in the perpendicular direction.

Traffic demand forecasting bridges from traffic flow theory to transportation planning. The classical **four-step model** projects future demand through trip generation (how many trips originate from each zone), trip distribution (which destinations those trips choose), mode choice (car, transit, walk, bike), and traffic assignment (which routes through the network each trip takes).

Each step produces inputs to the next, and the final traffic assignment provides the link-by-link flows that inform capacity analysis and facility design. Contemporary practice supplements the four-step model with activity-based models that track individual travelers through their daily schedules and increasingly with real-time data from GPS probes, cellular networks, and connected vehicles that allow continuous calibration against observed conditions.

> **Should transportation systems prioritize efficiency or resilience, and how does that choice shape long-term urban development?**

Chapter 7: Water Resources and Hydraulic Engineering

Water is the one natural resource that drives more civil engineering effort than any other. Cities need it delivered to every faucet in clean and pressurized form. They need it collected again after use and treated before discharge. They need rainfall directed off streets and roofs so buildings don't flood. They need rivers trained, dammed, and diverted so floods don't ruin downstream communities and droughts don't empty reservoirs. And they need all of this to happen during storms that vary from a gentle afternoon shower to a hundred-year cloudburst. Water resources engineering is the branch of civil engineering responsible for every one of those systems, from the raindrop hitting the ground through to the treated effluent returning to a river.

Two physical truths organize the discipline. Water flows downhill, gravity never changes, and given a path to lower elevation, water will find it. And water acts on surfaces with pressure that grows with depth, making every container, every pipe, and every dam a structural problem as well as a hydraulic one. Together these facts produce the design problems this chapter covers: analyzing watersheds to predict how much water arrives at a given point; computing how fast water moves through channels and pipes; designing dams and spillways that safely release excess water; and managing stormwater so that human development doesn't overwhelm natural drainage.

The Hydrologic Cycle and Watershed Analysis

The **hydrologic cycle** describes the continuous movement of water between the atmosphere, land surface, subsurface, and oceans. Water evaporates from oceans, lakes, and moist soil; it transpires from plants; it condenses into clouds; it precipitates as rain, snow, sleet, or hail; it runs off the surface toward streams and rivers; it infiltrates into the ground where it becomes soil moisture or groundwater; and it eventually returns to the oceans or the atmosphere to repeat the cycle. Every water resources problem is an application of this cycle to a particular place and moment.

Precipitation is the input that matters most to engineering design. Total annual precipitation varies from a few inches in desert regions to more than 200 inches in the rainiest parts of the Pacific Northwest and tropical coasts. But annual totals matter less to design than the intensity, duration, and frequency of individual storms.

Storms are characterized by **intensity-duration-frequency** (IDF) curves, plots that show the expected rainfall intensity for storms of various durations and return periods at a specific location. A 10-year 1-hour storm at a given site might deliver rainfall at 2 inches per hour for that hour. The same site might see 3.5 inches per hour in a 100-year 1-hour storm. Engineers choose the return period based on the consequences of being overwhelmed: a roadway culvert might be designed for a 50-

year storm, while a highway bridge over a major river typically is designed for 100-year or 500-year conditions.

Precipitation gages measure actual rainfall at point locations. **Radar-based estimates** cover larger areas using the Doppler shift of returning signals to estimate drop size and fall rate. **Satellite remote sensing** extends coverage globally, including over oceans and remote terrain where gages don't exist. Combining these sources produces the precipitation records that inform both real-time flood forecasting and long-term statistical design analyses.

A **watershed** (or drainage basin, or catchment) is all the land area from which precipitation drains to a single outlet point on a stream. Every point on a stream has its own upstream watershed that funnels water to it. Watersheds are delineated by tracing ridge lines on topographic maps or by using digital elevation models in geographic information systems. The watershed area, shape, slope, soil characteristics, vegetation cover, and existing development together determine how much runoff reaches the outlet for any given storm.

Runoff is the fraction of precipitation that actually reaches the stream. The rest infiltrates into soil, evaporates, or is intercepted by vegetation. **Runoff coefficient** (C) in the Rational Method is the simplest characterization: C equals the ratio of peak runoff rate to peak rainfall intensity times the watershed area. Values range from 0.10 for forest on sandy soil to 0.95 for pavement and rooftops.

The **Rational Formula**, $Q = CiA$, gives the peak discharge Q in cubic feet per second when C is the runoff coefficient, i is the rainfall intensity in inches per hour, and A is the watershed area in acres. It applies best to small urban watersheds up to about 200 acres, and it gives a useful preliminary estimate even where more sophisticated methods are required for final design.

The **SCS Curve Number method**, developed by what's now the Natural Resources Conservation Service, handles larger watersheds and more complex storms. It computes a runoff depth from a rainfall depth using a curve number (CN) that reflects the watershed's soil type, land cover, and antecedent moisture condition. Pavement has CN values near 98, meaning nearly all rainfall becomes runoff. Forests on well-drained sandy soils may have CN values as low as 30. Rural watersheds typically run 55 to 80. The method converts rainfall into runoff and then, through unit hydrograph techniques, converts runoff into a hydrograph of discharge versus time at the watershed outlet.

A **hydrograph** is the time history of discharge at a specific point on a stream. It rises as runoff from a storm reaches the point, peaks at some time after peak rainfall, and recedes as the watershed drains. Hydrograph shape depends on watershed characteristics: long narrow watersheds produce broad low peaks, while compact round watersheds produce sharp high peaks. The **time of concentration**, typically the time required for runoff to travel from the hydraulically most distant point in the watershed to the outlet, sets the critical storm duration for peak flow: storms shorter than the time of concentration don't fill the whole watershed; longer storms begin to see runoff decline as rainfall intensity drops.

Groundwater occupies a separate but connected compartment of the hydrologic cycle. Water that infiltrates past the root zone continues downward through the unsaturated zone until it reaches the water table, the upper surface of a saturated zone called an **aquifer**. Aquifers may be unconfined, exposed to atmospheric pressure at the water table, or confined between impermeable layers and under pressure.

Darcy's law relates groundwater flow velocity to the hydraulic gradient and the aquifer's hydraulic conductivity: $v = K \times i$, where v is apparent velocity, K is hydraulic conductivity in length per time, and i is the dimensionless hydraulic gradient. Wells pumped from aquifers draw down the water table in a cone of depression, and well interference, long-term groundwater depletion, and contamination all follow from the physics of Darcy flow applied to real geology.

Open Channel Flow and Manning's Equation

An **open channel** is any conduit where water flows with a free surface exposed to atmospheric pressure. Rivers, streams, canals, storm sewers running partially full, and even the gutters along streets are open channels. The physics differs fundamentally from pressure-pipe flow because the free surface can rise or fall in response to changes in discharge or channel geometry, and this interaction between flow and geometry produces the characteristic behaviors of open channels.

Open channel flow is classified along several dimensions. **Steady flow** occurs when discharge at a given cross-section doesn't change with time; **unsteady flow** does. **Uniform flow** has the same depth and velocity along the channel; **non-uniform flow** varies with position. **Subcritical flow** occurs when flow velocity is below the speed of wave propagation in the channel, and disturbances can travel upstream. **Supercritical flow** occurs when velocity exceeds wave speed, and disturbances only travel downstream.

The **Froude number**, $Fr = V / \text{sqrt}(g \times D)$, where V is velocity, g is gravitational acceleration, and D is hydraulic depth, determines which regime applies. Fr less than 1 is subcritical; Fr greater than 1 is supercritical; Fr equal to 1 is critical.

The difference between subcritical and supercritical flow is not just academic. Subcritical flow is tranquil and deep relative to its velocity, typical of rivers and large channels under normal conditions. Supercritical flow is shallow and fast, typical of spillways, steep gutters, and flood flows on steep streams. A **hydraulic jump** forms where supercritical flow transitions to subcritical, characterized by a turbulent standing wave that dissipates energy dramatically, useful for slowing water at the base of spillways but damaging if it forms where it wasn't intended.

Manning's equation is the workhorse formula for uniform flow in open channels, relating discharge to channel geometry, slope, and roughness. The equation in U.S. units is $V = (1.49 / n) \times R^{(2/3)} \times S^{(1/2)}$, where V is velocity in feet per second, n is Manning's roughness coefficient, R is the hydraulic radius (cross-sectional area divided by wetted perimeter) in feet, and S is the dimensionless channel slope. Multiplying V by the cross-sectional area A gives discharge Q. In SI units the constant 1.49 is replaced by 1.0.

Manning's n values are tabulated for every common channel surface. A smooth finished concrete channel has n around 0.012. An earthen channel with clean sides and bottom has n around 0.022. A natural stream with irregular cross-section, some brush, and occasional boulders has n around 0.035 to 0.050. Heavily vegetated floodplains push n toward 0.10 or higher. The selection of n is one of the few remaining judgment calls in hydraulics practice, and experience matters because small changes in n produce substantial changes in computed depth and velocity.

Manning's equation gives **normal depth**, the depth at which flow would uniformly occur for a given discharge and channel. For a given channel, **specific energy**, $E = y + V^2/(2g)$, where y is depth, describes the total mechanical energy in the flow relative to the channel bottom. Plotting specific energy against depth for a given discharge produces a curve with a minimum at the **critical depth**, the depth at which Froude number equals 1 and specific energy is minimized. Let's look at an example:

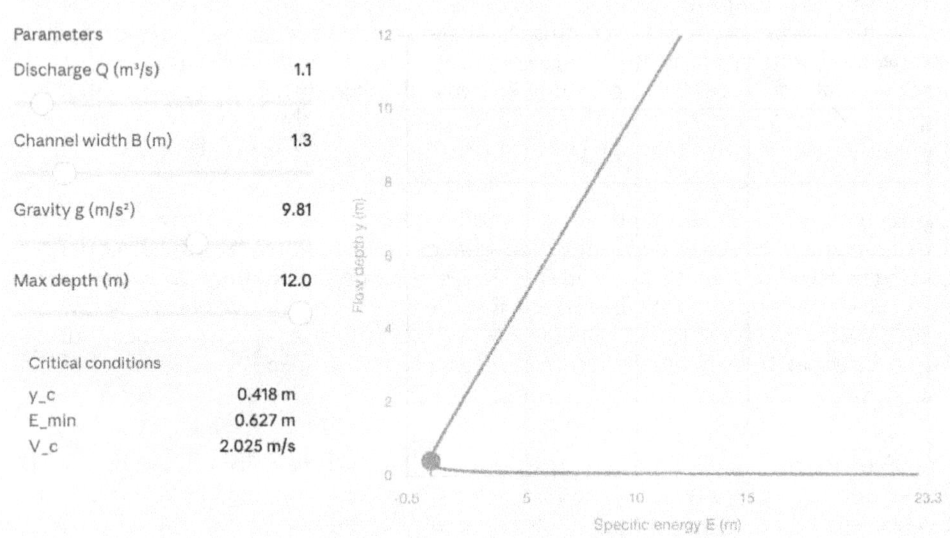

The curve shows **specific energy** $E = y + V^2/2g$ plotted against **flow depth** y for a rectangular channel, where $V = Q/(By)$.

A few things to note:

- The **dot** at the bottom of the "L" marks critical depth y_c, where the Froude number Fr = 1 and specific energy is minimized. For a rectangular channel, $y_c = (q^2/g)^{(1/3)}$ where $q = Q/B$ is the unit discharge.
- The **upper limb** of the curve is subcritical flow ($Fr < 1$, deep and slow), while the **lower limb** is supercritical ($Fr > 1$, shallow and fast) — two different depths can share the same specific energy (called alternate depths).
- Increasing **Q** shifts the entire curve to the right, raising the minimum energy needed to pass that discharge.
- Increasing **B** widens the channel, reducing unit discharge and shifting the curve left toward lower energies.

Two depths exist for every specific energy above the minimum: a smaller supercritical depth and a larger subcritical depth. This dual-depth behavior creates the characteristic shapes of backwater curves in rivers, drop-down curves near spillway crests, and the hydraulic jump wherever supercritical flow must become subcritical.

Critical flow conditions let engineers measure discharge with accuracy because the relationship between depth and discharge is uniquely defined at critical flow. **Flumes** and **weirs** built to force flow through critical depth are standard discharge measurement devices. A **broad-crested weir** raises the channel bottom enough that upstream flow must accelerate over the weir at critical depth. A **Parshall flume** narrows the channel through a specific geometry that ensures critical flow in the throat. Both let an operator measure the upstream water level and compute discharge directly from calibration curves without needing velocity measurements.

Gradually varied flow describes the smooth transitions between normal depth and imposed boundary conditions. Backwater from a dam extends upstream many miles in a large river, with depths gradually approaching normal depth as the distance from the dam grows. The governing equation combines Manning's equation with the energy equation to produce the **standard step method**, computing water surface profiles step by step from one cross-section to the next. The method is used in HEC-RAS, the U.S. Army Corps of Engineers software that dominates river hydraulic analysis for flood studies, floodplain mapping, and bridge design.

Sediment transport is the other side of open channel hydraulics. Flowing water picks up, moves, and deposits solid particles from the channel bed and banks. **Bed load** travels along the bottom by rolling, sliding, and saltating. **Suspended load** moves through the water column, held aloft by turbulence. Shields' parameter predicts the threshold velocity required to initiate motion of a particle of given size and density. Above the threshold, sediment transport rates rise rapidly with velocity, and channels adjust their geometry and slope over time to carry the sediment supplied from upstream. Hydraulic structures must account for sediment to avoid siltation upstream of dams, scour at bridge piers, and degradation of channels downstream of reservoirs where the sediment load has been trapped.

Pipe Networks, Pressure Systems, and Pump Design

Water distribution systems, wastewater collection systems, and industrial process piping all involve **pressure pipe flow**, where the pipe is completely full and the water is under pressure greater than atmospheric. The analysis differs from open channel flow because the cross-section is fixed and the pressure varies along the pipe in response to friction, elevation changes, and pumps.

The **Bernoulli equation** is the foundation of every pipe flow calculation. Written between two points along a streamline, it states that the sum of pressure head (p/γ), velocity head ($V^2/2g$), and elevation head (z) at one point equals the same sum at another point, adjusted for friction losses and pump or turbine work between them: $p1/\gamma + V1^2/2g + z1 + h\_pump = p2/\gamma + V2^2/2g + z2 + h\_friction + h\_minor$. Each term has units of length (feet of water), and together they express conservation of energy along the flow path.

Friction losses in full pipes are computed using the **Darcy-Weisbach equation**, h_f = f × (L/D) × V²/(2g), where f is the Darcy friction factor, L is pipe length, D is diameter, V is velocity, and g is gravitational acceleration. The friction factor f depends on the Reynolds number (dimensionless ratio of inertial to viscous forces) and the relative roughness (pipe roughness divided by diameter). The **Moody diagram**, a log-log plot developed in the 1940s (below), gives f graphically for any combination of Reynolds number and relative roughness.

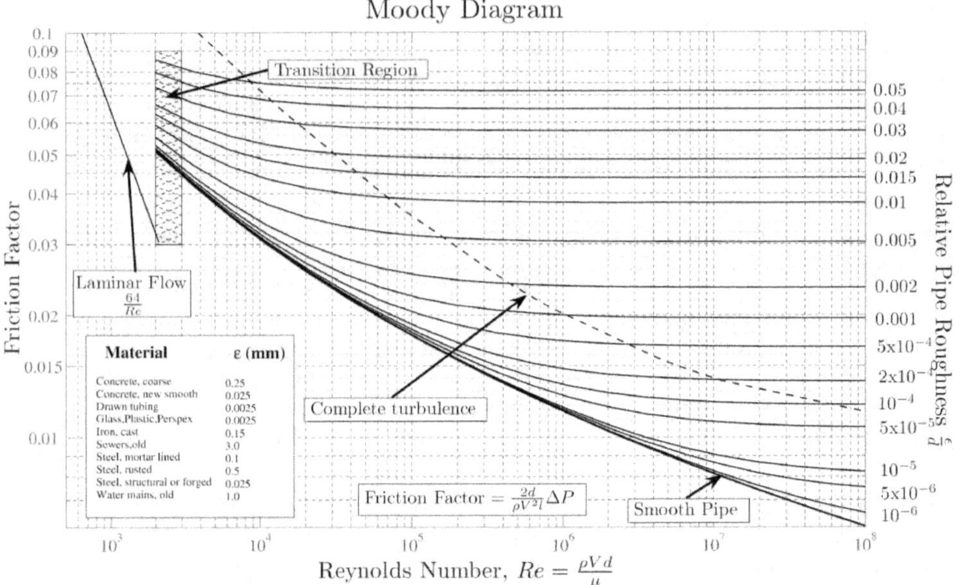

The **Colebrook-White equation** provides the same information analytically, though it's implicit and requires iteration to solve.

For water supply calculations at moderate velocities and standard pipe materials, the empirical **Hazen-Williams equation** offers a simpler alternative: h_f = 10.44 × L × Q^1.85 / (C^1.85 × D^4.87) in U.S. units, where C is the Hazen-Williams coefficient, Q is flow in gpm, and D is diameter in inches. C values range from 100 for old cast iron pipe to 140 for new plastic or cement-lined ductile iron. Hazen-Williams is less accurate than Darcy-Weisbach at very high or very low velocities but is convenient for the range typical of municipal water systems.

Minor losses occur at fittings, valves, bends, expansions, contractions, entrances, and exits. Each device has a characteristic loss coefficient K such that the head loss equals K × V²/(2g). A sharp-edged entrance has K = 0.5. A long-radius 90-degree elbow has K = 0.3. A fully open gate valve has K close to 0. Partially closed valves can have K values in the hundreds. In long pipelines, minor losses are indeed minor and often neglected; in short lines with many fittings, they can dominate the total head loss.

Pipe networks connect multiple sources to multiple demands through a mesh of interconnected pipes. Water distribution systems in cities are typical networks, with treated water entering at pump stations or storage tanks and flowing through mains, submains, and distribution pipes to every service connection. Network analysis must solve for the flow in every pipe and the pressure at every junction simultaneously. Two conservation laws constrain the system: **continuity** at each junction (total flow in equals total flow out plus any demand) and **energy** around each loop (sum of head losses around a closed loop equals zero).

For small networks, these equations can be solved by the **Hardy Cross method**, an iterative procedure where flows are assumed, loop corrections are calculated, and the process repeats until the corrections become negligibly small. The Hardy Cross method, developed in 1936, dominated water system design for decades. Modern practice uses commercial software like EPANET, WaterCAD, or InfoWater, which solves the full system of equations by matrix methods and handles large networks with thousands of pipes and junctions quickly. These tools also simulate transient conditions, water quality parameters, pressure-dependent demands, and fire flow requirements that static analysis can't address.

Pumps add energy to water to raise it against gravity or push it through friction. The two main types used in civil water systems are centrifugal pumps for high flow at moderate head and positive-displacement pumps for low flow at high head or for slurries. **Centrifugal pumps** have an impeller that rotates inside a casing, flinging water outward by centrifugal force and converting velocity to pressure in the volute. Their performance is described by **pump curves** that plot head against flow at a given rotational speed. A specific pump's curve intersects the **system curve** (the head required to push flow through the piping system) at the **operating point**, the flow and head at which the pump will actually deliver water.

Pump selection requires matching the system demands to the available pump curves. Pumps operating far from their **best efficiency point** (BEP) waste energy and wear out quickly. Pumps run in parallel double the flow at the same head; pumps run in series double the head at the same flow. **Variable-speed drives** allow the pump speed to adjust continuously, matching output to demand in real time and saving substantial energy in systems with fluctuating loads.

Cavitation is the destructive phenomenon that occurs when water pressure inside a pump drops below the vapor pressure, causing bubbles to form and then collapse violently as they move into higher-pressure regions of the pump. The damage looks like pitting on the impeller surfaces and degrades pump performance rapidly. Cavitation is prevented by ensuring adequate **Net Positive Suction Head** (NPSH) at the pump inlet: NPSH available (a function of the suction piping, liquid temperature, and atmospheric pressure) must exceed NPSH required (a pump characteristic provided by the manufacturer). Violating NPSH requirements is one of the most common causes of pump failure, and every pump installation includes an explicit NPSH check.

Water hammer is the pressure surge that propagates through a pipe when flow changes suddenly, as when a valve closes quickly or a pump trips off. The magnitude depends on the change in velocity and the pipe's wave celerity, typically 3,000 to 4,000 feet per second in metal pipes. A rapid valve closure can generate pressure rises of hundreds of psi, enough to burst pipes that operate safely at steady pressures.

Water hammer is managed by slow-closing valves, surge tanks, air chambers, and careful pump startup and shutdown sequences designed to limit rates of change.

Dams, Spillways, and Reservoir Operation

A **dam** is a barrier that stops or slows the flow of water in a river, impounding the water behind it to form a **reservoir**. Dams serve multiple purposes: water supply, irrigation, hydroelectric power generation, flood control, recreation, navigation, and environmental flow management. Large dams typically serve several of these functions simultaneously, and the operation of the reservoir balances competing demands on the stored water.

Dams fall into structural categories based on how they resist the water pressure acting against them. **Gravity dams** are massive structures, usually of concrete, that resist overturning and sliding through their own weight. They work for any height but become uneconomical at very great heights because concrete volume scales with height squared. **Arch dams** transmit water pressure into the canyon walls through arch action, allowing a thinner and lighter structure. They work only in narrow canyons with competent rock abutments. **Buttress dams** use a watertight upstream face supported by a series of buttresses on the downstream side, reducing concrete volume compared to a gravity dam. **Embankment dams** are constructed from compacted earth or rockfill and use the weight of the embankment and its low permeability or internal impermeable core to resist water pressure and prevent seepage failure.

Every dam fails in specific modes that the design must prevent. **Overtopping** occurs when water rises above the dam crest during a flood that exceeds the spillway capacity. Overtopping is catastrophic for embankment dams, where flowing water rapidly erodes the downstream slope, and it's a concern even for concrete dams if it continues long enough to undermine the foundation. **Seepage failure** in embankment dams results from water migrating through the dam body or its foundation at velocities high enough to carry soil particles with it, creating internal erosion or piping that progressively enlarges until the dam fails suddenly. **Sliding** and **overturning** failures in concrete dams are structural failures of the entire monolith against its foundation. **Slope failure** of the downstream or upstream face can reduce the dam's cross-section below the margin needed for stability.

Spillways release water from the reservoir when inflows exceed storage capacity, preventing overtopping of the dam. The spillway is the most important safety feature of any dam, and its capacity is sized to pass the **probable maximum flood** (PMF) for high-hazard dams or smaller design floods for lower-hazard structures. The PMF is an estimate of the largest flood that could physically occur at a site given the most severe meteorological conditions. It's used for dams whose failure would cause loss of life or major economic damage, while smaller dams might design spillways for the 100-year or 1,000-year flood.

Spillways take several forms. **Ogee spillways** have a crest shaped to match the underside of a nappe of water falling freely over a sharp weir, producing efficient discharge without negative pressures on the spillway surface. **Side-channel spillways** collect water along a long crest and turn it 90 degrees into a channel parallel to the dam. **Chute spillways** convey water from the crest down a steep inclined channel to a stilling basin. **Tunnel spillways** carry water through the abutment in a steep tunnel. **Morning glory** or **glory hole spillways** drop water into a vertical shaft through a horizontal bell-shaped inlet, useful where space constraints preclude a conventional spillway. Every spillway terminates in some form of **energy dissipator**, typically a **stilling basin** sized to force a hydraulic jump that destroys the high-velocity flow's kinetic energy before it reaches the downstream channel.

Outlet works are smaller conduits that release water from the reservoir for normal operations, irrigation deliveries, downstream releases, or emergency drawdown. They typically include control gates or valves, conveyance tunnels or pipes through the dam, and downstream dissipators. Unlike spillways, which operate only during floods, outlet works operate continuously or frequently throughout the year.

Reservoir operation balances storage against downstream needs and upstream inflows. A reservoir has a total capacity divided into zones with different operational purposes. **Dead storage** lies below the lowest outlet and cannot be drawn down; it accumulates sediment over the life of the reservoir. **Conservation storage** holds water for water supply, irrigation, and baseline environmental flows. **Flood control storage** is deliberately kept empty so that the reservoir can absorb incoming floods. **Surcharge storage** is the zone between the normal maximum water surface and the top of the dam, filled only during severe floods when the spillway is already discharging.

Operators manage the reservoir through **rule curves** that specify target elevations by season, storage allocations for different purposes, and release schedules coordinated with downstream requirements. Spring reservoir elevations are typically drawn down to create flood control storage before the snowmelt or rainy season. Summer and fall elevations rise as water is captured for later use. The operational challenge is balancing conflicting demands: holding water for summer supply risks inadequate flood storage if an unseasonal storm arrives; releasing water for flood protection risks shortages later in the year if the anticipated flood doesn't materialize.

Stormwater Management and Flood Control Structures

Urban development replaces pervious soils and vegetation with pavements, rooftops, and compacted fill. Runoff volumes and peak flows from developed watersheds increase, sometimes dramatically. A typical suburban subdivision generates three to five times the peak runoff of the forest or farmland it replaced. Stormwater management designs the systems that handle this extra flow, protect water quality, and prevent flooding.

Storm sewers are the pipe networks that convey collected runoff from streets and roofs to discharge points in receiving streams or detention facilities. They typically run as separate systems parallel to (but independent of) sanitary sewers in newer cities, though older cities often have **combined sewers** that carry both stormwater and sanitary flow to treatment plants, overflowing to receiving waters during large storms. Storm sewer design applies open channel flow concepts to partially full pipes, sizing each reach to pass the design flow without surcharging above the pipe crown and without creating velocities too high (which erode the pipe) or too low (which allow sediment to deposit). Manning's equation applies directly with an appropriate n for concrete or corrugated metal pipe.

Inlets and **catch basins** collect runoff from streets and direct it into storm sewers. Grate inlets, curb-opening inlets, and combination inlets each have capacity characteristics that depend on street slope, gutter geometry, and flow rate. Inlets are spaced to prevent water from ponding across traffic lanes or backing up onto sidewalks during design storms. **Inlet spread**, the width of flow across a travel lane, is limited by code to protect driver safety: typically no wider than 4 to 8 feet for interstate highways and wider for lower-speed streets.

Detention and **retention** facilities manage peak flows and volumes before releasing water to downstream systems. **Detention basins** temporarily store runoff and release it slowly, reducing the peak discharge to pre-development levels even though the total runoff volume increases. **Retention basins** permanently capture runoff, infiltrating it or letting it evaporate, reducing both peak and volume. Wet retention ponds hold a permanent pool that provides water quality treatment through sedimentation and biological processes. Dry detention basins drain completely after each storm.

Green infrastructure increasingly complements or replaces traditional gray infrastructure for stormwater management. **Bioretention cells** (rain gardens) capture runoff in a shallow depression filled with engineered soil and vegetation that filters pollutants and infiltrates water. **Permeable pavements** allow water to pass through the pavement structure to underlying stone storage and eventual infiltration. **Green roofs** capture and evapotranspire rooftop runoff. **Bioswales** are vegetated channels that slow and filter flows while conveying them. Together these techniques mimic the hydrologic behavior of natural vegetation, reducing the downstream loading on storm sewers and receiving waters.

Flood control structures operate at a larger scale than urban stormwater. **Levees** are earth or concrete walls built along riverbanks to contain floodwaters within the channel. **Floodwalls** are reinforced concrete versions of the same idea where right-of-way is limited. **Flood bypasses** are channels that divert excess flow around communities to safer release points downstream. **Reservoirs** dedicated partly or wholly to flood control store incoming floods temporarily and release them gradually at rates the downstream channel can safely carry. The U.S. Army Corps of Engineers operates a network of flood control reservoirs across the country, often in combination with hydroelectric and water supply functions.

Floodplain mapping identifies the areas subject to flooding at various return periods. The National Flood Insurance Program uses 100-year floodplain maps to regulate development and determine insurance requirements. **Base flood elevations** establish the level that floor slabs of insurable structures must exceed. **Floodways** are the portions of the floodplain that carry most of the flood flow; development within them is severely restricted because obstructions would raise flood elevations upstream. **Flood fringes** outside the floodway can be developed if structures are elevated above the base flood elevation.

Bridges and **culverts** cross waterways and must be designed to pass design floods without unacceptable damage. Bridge waterway openings must be sized large enough that the backwater upstream of the bridge doesn't exceed allowable limits, typically 1 foot for most sites. Scour at bridge piers and abutments must be estimated using empirical formulas and the foundations designed to extend below the maximum scour depth. Culverts must pass design flows without overtopping the roadway, and their hydraulic behavior is analyzed using different methods depending on whether they flow under inlet control (capacity limited by the culvert entrance geometry) or outlet control (capacity limited by the barrel friction and tailwater conditions downstream).

Every element of this chain, from raindrops falling in the watershed to culverts passing flood flows beneath highways, follows from the same hydraulic principles applied at different scales and boundary conditions. Mastering those principles is what lets the water resources engineer move from the scale of a rooftop gutter to the scale of a major river system without changing the fundamental approach to the problem.

How should engineers design water systems when historical data is no longer a reliable predictor of future conditions?

Chapter 8: Environmental and Sanitary Engineering

Clean water coming in, dirty water going out, solid waste disposed of safely, and air that doesn't poison the people breathing it. Environmental and sanitary engineering handles all four, and each one touches every person in every community every day. The discipline sits at the intersection of chemistry, biology, fluid mechanics, and public health, applying engineering methods to problems that are fundamentally about keeping human populations alive and healthy in dense concentrations. Before cities had sanitary engineering, they had cholera, typhoid, and infant mortality rates that would horrify a modern observer. The infrastructure this chapter describes is the reason those diseases are historical footnotes in the developed world rather than daily realities.

The scope has expanded well beyond its nineteenth-century origins in water purification and sewage disposal. Modern environmental engineering encompasses hazardous waste remediation, air pollution control, industrial process emissions, noise abatement, and climate-related infrastructure adaptation. But the core remains water and waste: getting clean water to people, collecting what they've used, treating it before it returns to the environment, and managing the solid byproducts of civilization so they don't contaminate land, water, or air. This chapter covers the essential systems in each domain.

Drinking Water Treatment Processes and Distribution

Surface water drawn from rivers, lakes, and reservoirs contains turbidity, organic matter, pathogens, dissolved minerals, and increasingly, trace pharmaceuticals and industrial chemicals. Groundwater from wells is often cleaner but may carry elevated iron, manganese, hardness, arsenic, or nitrate depending on the local geology. Neither source is safe to drink without treatment, and the treatment train must reduce every contaminant to levels that meet regulatory standards, primarily the Safe Drinking Water Act and its implementing regulations in the United States, or equivalent national standards elsewhere.

A conventional surface water treatment plant follows a sequence that hasn't changed fundamentally in over a century, though every step has been refined. Raw water enters the plant and receives a **coagulant**, typically aluminum sulfate (alum) or ferric chloride, dosed at concentrations determined by jar testing. The coagulant destabilizes suspended particles by neutralizing the electrical charges that keep them apart in suspension. **Rapid mixing** distributes the coagulant uniformly through the raw water in seconds.

The water then enters **flocculation basins**, where slow, gentle mixing brings the destabilized particles into contact with each other so they aggregate into larger clumps called **floc**. Flocculation takes 20 to 40 minutes at controlled velocity gradients. Too much mixing breaks the floc apart; too little fails to build it large enough to settle.

From flocculation, the water moves to **sedimentation basins** (or clarifiers), large tanks where flow velocity drops low enough that the floc settles to the bottom under gravity. Residence time is typically 2 to 4 hours. The settled sludge is collected from the basin floor and sent to sludge processing. The clarified water, now substantially reduced in turbidity and particulate matter, flows over weirs at the basin outlet.

Filtration follows sedimentation. Rapid sand filters or dual-media filters (anthracite coal over sand) remove the remaining suspended particles that didn't settle. Water percolates downward through the filter media, and particles are captured by attachment to the media grains. Filters gradually clog and must be backwashed periodically by reversing the flow to flush out accumulated solids. A well-operated filter reduces turbidity to below 0.3 NTU (nephelometric turbidity units), which is critical because low turbidity is a prerequisite for effective disinfection.

Disinfection is the final and most important barrier against waterborne disease. Chlorine remains the most widely used disinfectant in municipal water treatment. It's effective against bacteria and most viruses at low concentrations, it's cheap, and it provides a **residual** that persists through the distribution system, continuing to protect the water from recontamination between the plant and the tap. Free chlorine residuals of 0.2 to 0.5 mg/L at the point of use are typical targets. **Chloramines**, formed by adding ammonia to chlorinated water, provide a more stable but less potent residual and are used in distribution systems where free chlorine would decay too quickly or produce excessive disinfection byproducts.

Disinfection byproducts (DBPs) are the unintended consequence of chlorine reacting with natural organic matter in the water. **Trihalomethanes** (THMs) and **haloacetic acids** (HAAs) are the two regulated classes, both associated with long-term health risks at elevated concentrations. Plants manage DBPs by reducing organic precursors before chlorination (through enhanced coagulation or activated carbon), by using alternative disinfectants like ozone or ultraviolet light for primary disinfection, and by switching to chloramines for the distribution residual. **Ozone** is a strong oxidant that kills pathogens quickly and produces no chlorinated byproducts, though it does form bromate in waters containing bromide. **Ultraviolet (UV) disinfection** inactivates pathogens by damaging their DNA and produces no chemical byproducts at all, though it provides no residual protection downstream.

Advanced treatment processes address specific contaminants that conventional treatment doesn't remove adequately. **Granular activated carbon** (GAC) adsorbs dissolved organic chemicals, taste and odor compounds, and some synthetic contaminants. **Membrane filtration** technologies, including microfiltration, ultrafiltration, nanofiltration, and reverse osmosis, physically separate contaminants by forcing water through semi-permeable membranes with progressively finer pore sizes. Reverse osmosis removes dissolved salts and is the basis of desalination plants that convert seawater to drinking water, though at high energy cost.

The treated water enters the **distribution system**, a pressurized pipe network that delivers water to every home, business, fire hydrant, and industrial user in the service area. Distribution systems maintain pressure between 40 and 80 psi at service connections, with higher pressures at the plant or pump station to overcome friction and elevation changes across the service area. **Storage tanks** and **elevated reservoirs** buffer demand fluctuations: they fill during low-demand periods (typically nighttime) and drain during peaks (morning and evening), reducing the required pump capacity and providing emergency reserves for fire suppression.

Water quality can degrade in the distribution system itself. Long residence times in dead-end mains allow the chlorine residual to decay and bacterial regrowth to occur. Corrosive water attacks pipe interiors, releasing lead from older service lines and solder joints, iron from unlined cast iron mains, and copper from household plumbing.

Corrosion control, typically through pH adjustment and addition of orthophosphate inhibitors that form a protective scale on pipe surfaces, is now a standard treatment requirement. The crisis in Flint, Michigan, demonstrated what happens when corrosion control fails: lead concentrations at the tap spiked to levels many times the action level, exposing thousands of residents to a preventable neurotoxin.

Wastewater Collection and Biological Treatment Systems

What goes in must come out. Every gallon of water delivered to a building emerges as wastewater carrying human waste, food particles, soap, detergents, pharmaceuticals, and whatever else goes down the drain. Collecting this flow and treating it before it reaches receiving waters is the other half of the sanitary engineering equation.

Sanitary sewers collect wastewater from buildings and convey it by gravity to treatment plants. The network consists of building laterals connecting to street mains, which join progressively larger trunk sewers and interceptors as flow accumulates. Pipe sizes range from 8-inch laterals to tunnels 20 feet or more in diameter for major interceptors in large cities. Grades are set to maintain self-cleansing velocities, typically at least 2 feet per second at design flow, to prevent solids from depositing and eventually blocking the pipe. Where gravity flow can't reach the treatment plant, **lift stations** (pump stations) raise the wastewater to a higher elevation and discharge it into the next gravity segment.

Wastewater arrives at the treatment plant as a dilute mixture that's roughly 99.9 percent water by weight. The remaining 0.1 percent contains suspended solids, dissolved organic matter measured as **biochemical oxygen demand** (BOD), nutrients (nitrogen and phosphorus), pathogens, and trace chemicals. BOD quantifies the oxygen that microorganisms would consume while decomposing the organic matter, and it's the principal measure of wastewater strength. Raw domestic wastewater typically has BOD around 200 to 300 mg/L and total suspended solids (TSS) around 200 to 250 mg/L.

Treatment proceeds through stages. **Preliminary treatment** removes large debris, rags, plastics, and grit that would damage downstream equipment. Bar screens intercept solids larger than about 1/4 inch. Grit chambers settle sand, gravel, and other heavy mineral particles. Neither step reduces BOD or TSS significantly, but both protect the biological processes that follow.

Primary treatment settles suspended solids in large sedimentation tanks similar in concept to those in water treatment but handling a much dirtier flow. Residence time runs 1.5 to 2.5 hours. Primary clarifiers remove roughly 50 to 70 percent of TSS and 25 to 40 percent of BOD. The settled material, called **primary sludge**, is thick and rich in organic matter and requires further processing.

Secondary treatment is the biological core of the plant. Microorganisms, primarily bacteria, consume the dissolved and colloidal organic matter remaining after primary treatment, converting it to cell mass and carbon dioxide. The engineer's job is to create conditions that let these organisms thrive: enough oxygen, enough contact time, the right temperature range, and a balance between the food supply (BOD) and the biomass available to consume it.

The **activated sludge process** is the most common secondary treatment system worldwide. Wastewater flows into an aeration basin where it mixes with a concentrated suspension of microorganisms called **mixed liquor**. Air or pure oxygen is injected through diffusers at the basin bottom, providing the dissolved oxygen the bacteria need. The bacteria metabolize the organic matter over a detention time of 4 to 8 hours. The mixed liquor then flows to a **secondary clarifier**, where the biological floc settles. The settled biomass, now called **activated sludge**, is returned to the aeration basin to maintain the microbial population. Excess sludge, produced as the bacterial population grows from consuming organic matter, is wasted from the system and sent to sludge processing.

Activated sludge systems come in many variants. **Conventional** activated sludge uses a plug-flow aeration basin. **Complete-mix** systems use a fully mixed reactor. **Extended aeration** operates at long solids retention times, producing less excess sludge but requiring larger basins. **Sequencing batch reactors** (SBRs) perform all steps, aeration, settling, decanting, in a single basin through timed cycles. **Oxidation ditches** use long looping channels with surface aerators, common in smaller communities. **Membrane bioreactors** (MBRs) replace the secondary clarifier with submerged membranes that filter the mixed liquor directly, producing an effluent clear enough to reuse without further filtration.

Nutrient removal has become a standard requirement for plants discharging to sensitive receiving waters. Excess nitrogen causes algal blooms and oxygen depletion in rivers and estuaries. Excess phosphorus drives eutrophication in lakes. **Biological nitrogen removal** exploits two microbial processes: **nitrification**, where autotrophic bacteria convert ammonia to nitrate in aerobic conditions, and **denitrification**, where heterotrophic bacteria convert nitrate to nitrogen gas in anoxic (oxygen-free) conditions. Alternating aerobic and anoxic zones within the aeration basin accomplishes both transformations in a single system. **Biological phosphorus removal** uses bacteria that accumulate phosphorus beyond their metabolic needs when cycled between anaerobic and aerobic conditions. Chemical precipitation with alum or ferric chloride supplements or replaces biological phosphorus removal where stricter limits apply.

Sludge processing handles the concentrated organic solids removed during treatment. Sludge is thickened (to reduce volume), stabilized (to reduce pathogen content and odor), dewatered (to reduce moisture for disposal), and ultimately disposed of by land application, landfilling, or incineration.

Anaerobic digestion is the most common stabilization method: sludge is held in sealed, heated tanks at 95 to 100 degrees Fahrenheit for 15 to 30 days while anaerobic bacteria decompose the volatile organic solids, producing **biogas** (roughly 60 percent methane and 40 percent carbon dioxide) as a byproduct. Many plants capture the biogas and use it to generate electricity or heat, offsetting a significant fraction of the plant's energy consumption. The digested sludge, now called **biosolids**, is dewatered to 20 to 30 percent solids by belt filter presses, centrifuges, or plate-and-frame presses and is applied to agricultural land as a soil amendment where regulations permit.

A well-operated secondary treatment plant produces an effluent with BOD and TSS each below 30 mg/L, meeting the national secondary treatment standard. Advanced plants targeting water reuse or discharge to sensitive environments achieve effluent BOD below 5 mg/L, TSS below 5 mg/L, and nutrient concentrations in the low single digits.

Solid Waste Management and Landfill Engineering

Every person in the United States generates roughly 4.5 pounds of municipal solid waste per day. Multiplied across 330 million people and 365 days, the total exceeds 290 million tons per year. Managing this volume so it doesn't contaminate water, pollute air, or create public health hazards is a major engineering enterprise.

The **waste management hierarchy** prioritizes reduction, reuse, recycling, energy recovery, and landfilling, in that order. Source reduction eliminates waste before it's created. Recycling recovers materials for remanufacture. Composting diverts organic waste from landfills and produces a useful soil product. **Waste-to-energy** facilities combust municipal solid waste at high temperatures, generating steam to drive turbines for electricity and reducing waste volume by 85 to 90 percent by weight. Modern waste-to-energy plants include extensive air pollution controls, scrubbers, fabric filters, and selective catalytic reduction, to meet emission standards. Despite these options, landfilling remains the disposal method for roughly half of all municipal solid waste in the United States.

A modern **sanitary landfill** is an engineered containment facility, not an open dump. Federal regulations under RCRA Subtitle D require a composite liner system on the bottom and sides, a leachate collection system, a final cover system, groundwater monitoring, gas management, and financial assurance for closure and post-closure care extending at least 30 years. The engineering challenge is containing the waste and its byproducts for decades after the landfill stops accepting material.

The **liner system** prevents leachate, the contaminated liquid that percolates through the waste, from reaching groundwater. A typical composite liner combines a geomembrane (60-mil high-density polyethylene, or HDPE) on top of a 2-foot layer of compacted clay with a hydraulic conductivity no greater than 1 × 10^-7 cm/sec.

Double liner systems, required for hazardous waste, add a second composite liner beneath the first with a leak detection layer between them.

Leachate forms as precipitation infiltrates through the waste and dissolves or suspends organic and inorganic contaminants. Its composition varies with waste age: young leachate from active cells is high in BOD, volatile fatty acids, and ammonia; mature leachate from older cells has lower BOD but elevated refractory organics and heavy metals. Leachate is collected by a network of perforated pipes embedded in a drainage layer above the liner, pumped to the surface, and either treated on-site or trucked to a wastewater treatment plant for co-treatment.

Landfill gas results from the anaerobic decomposition of organic waste buried below the surface. The gas is roughly 50 percent methane and 50 percent carbon dioxide, with traces of hydrogen sulfide, volatile organic compounds, and other gases. Methane is both a potent greenhouse gas and an explosion hazard if it migrates into enclosed spaces. Federal regulations require gas collection and control at landfills exceeding size and emission thresholds. Many landfills flare the collected gas; increasingly, they clean and upgrade it to pipeline-quality renewable natural gas or use it to fuel generators, producing electricity sold to the grid.

Daily operations at a landfill involve spreading incoming waste in thin lifts, compacting it with heavy steel-wheeled compactors, and covering each day's waste with 6 inches of soil or alternative daily cover (tarps, foam, mulch). Compaction densities of 1,200 to 1,500 pounds per cubic yard maximize the useful life of the airspace. **Cell design** organizes the landfill into discrete filling areas that can be opened, filled, and closed sequentially, minimizing the active face exposed to precipitation and wind.

The **final cover** system placed over the completed landfill includes a low-permeability barrier (geomembrane, compacted clay, or both) to limit infiltration, a drainage layer to remove water that gets through the barrier, a protective soil layer, and a vegetative surface layer. The cover must accommodate differential settlement as waste decomposes and consolidates beneath it, sometimes losing 20 to 30 percent of its original height over decades. Cover integrity is critical: a breached cover allows water infiltration that generates more leachate and can reactivate decomposition in ways the gas collection system must handle.

Air Quality Control and Emissions Regulation

Air pollution differs from water and solid waste problems in one fundamental way: air is shared across boundaries without any collection system. A factory smokestack in one state sends its emissions downwind to another. A wildfire in one country degrades air quality across an ocean. This transboundary nature makes air quality both a local engineering challenge and a regulatory problem that extends from municipal zoning to international agreements.

The **Clean Air Act**, first enacted in 1963 and significantly amended in 1970 and 1990, is the foundational U.S. legislation. It establishes **National Ambient Air Quality Standards** (NAAQS) for six **criteria pollutants**: particulate matter (PM10 and PM2.5), sulfur dioxide (SO2), nitrogen dioxide (NO2), carbon monoxide

(CO), ozone (O3), and lead (Pb). Primary standards protect public health with an adequate margin of safety. Secondary standards protect public welfare, including visibility, vegetation, and building materials. Areas that fail to meet the standards are designated **nonattainment areas** and must implement plans to reach compliance.

Each criteria pollutant has distinct sources and health effects. **Particulate matter** originates from combustion processes, road dust, construction, agriculture, and natural sources like wildfires. Fine particles (PM2.5, smaller than 2.5 micrometers) penetrate deep into the lungs and are associated with cardiovascular and respiratory disease.

Sulfur dioxide comes primarily from coal and heavy fuel oil combustion and causes respiratory irritation and acid deposition. **Nitrogen oxides** (NO and NO2, collectively NOx) form during high-temperature combustion in vehicles and power plants and contribute to smog and ground-level ozone formation. **Ozone** at ground level is a secondary pollutant formed by photochemical reactions between NOx and volatile organic compounds (VOCs) in sunlight, making it a summer problem in urban areas. **Carbon monoxide** comes almost entirely from incomplete combustion in vehicle engines. **Lead** emissions have dropped more than 99 percent since the phase-out of leaded gasoline, but industrial sources like smelters and battery recyclers still require monitoring.

Emission controls at stationary sources reduce pollutant releases from power plants, refineries, factories, and other industrial facilities. **Electrostatic precipitators** (ESPs) remove particulate matter by charging particles as they pass through an electric field and collecting them on grounded plates. Collection efficiencies exceed 99 percent for well-designed and maintained units. **Fabric filters** (baghouses) pass exhaust gas through tightly woven bags that trap particles on their surface, achieving similar efficiencies with better performance on fine particles. **Wet scrubbers** contact exhaust gas with a liquid spray that captures both particles and soluble gases like SO2. **Flue gas desulfurization** (FGD) systems, most commonly lime or limestone scrubbers, remove 90 to 98 percent of SO2 from coal-fired power plant exhaust. **Selective catalytic reduction** (SCR) injects ammonia or urea into the exhaust stream ahead of a catalyst bed that converts NOx to nitrogen gas and water vapor.

Mobile source controls address vehicle emissions through a combination of engine design, fuel quality, and exhaust aftertreatment. **Catalytic converters** on gasoline vehicles oxidize CO and hydrocarbons to CO2 and water and reduce NOx to nitrogen gas, achieving reductions above 90 percent when the catalyst is at operating temperature. **Diesel particulate filters** (DPFs) capture soot from diesel exhaust and burn it off periodically through regeneration. **Selective catalytic reduction** on heavy-duty diesel engines reduces NOx using a urea-based reductant (diesel exhaust fluid). Fuel standards limiting sulfur content have been essential to enabling catalytic aftertreatment, because sulfur poisons catalysts and must be kept below 15 ppm in highway diesel and 10 ppm in gasoline.

Air quality modeling predicts how emissions from specific sources disperse through the atmosphere and affect concentrations at ground level. **Gaussian dispersion models** like AERMOD estimate pollutant concentrations downwind of point, area, and volume sources as functions of emission rate, wind speed and direction, atmospheric stability, stack height, and terrain. Modeling results inform permit decisions, determine required stack heights, and evaluate whether a proposed facility will cause or contribute to a NAAQS violation. **Photochemical models** like CMAQ simulate the complex chemistry of ozone and secondary particulate formation across regional domains, informing state implementation plans for ozone nonattainment areas.

Indoor air quality has received increasing attention as people spend 85 to 90 percent of their time indoors. Sources include building materials that off-gas formaldehyde and VOCs, combustion appliances that release CO and NOx, mold growth from moisture intrusion, radon migrating from soils into basements, and environmental tobacco smoke. Ventilation standards, published by ASHRAE, specify minimum outdoor air supply rates by occupancy type to dilute indoor contaminants. Tighter building envelopes built for energy efficiency must be paired with adequate mechanical ventilation to prevent indoor air quality from degrading, a balance that demands coordination between the building's environmental, mechanical, and architectural engineers.

> **Where should the line be drawn between engineering solutions and policy decisions in protecting environmental and public health?**

Chapter 9: Construction Management and Project Delivery

Design drawings and specifications describe a structure that doesn't exist yet. Construction management is the discipline that turns those documents into a physical reality, on time, within budget, and without killing anyone in the process. It sits at the boundary between engineering and business, demanding fluency in both. A construction manager reads structural drawings, understands soil reports, and negotiates subcontracts in the same afternoon. The work is sequential, interdependent, weather-sensitive, labor-intensive, and unforgiving of mistakes that get buried under the next pour of concrete.

Construction is also the phase where most of the money gets spent. Design fees typically run 5 to 15 percent of total project cost. The remaining 85 to 95 percent flows through construction, paying for materials, labor, equipment, subcontractors, insurance, and the overhead of managing all of it. Getting the construction phase right, or wrong, determines whether the project's owner receives good value or watches costs spiral beyond what the finished facility is worth.

Estimating, Scheduling, and Critical Path Method

Before a contractor can bid a project or an owner can set a budget, someone has to estimate what the work will cost. **Cost estimating** progresses through levels of detail that match the project's stage of development. A conceptual estimate at the planning phase might use cost per square foot based on similar completed buildings. A detailed estimate at the bid stage prices every item of work individually: cubic yards of concrete, tons of reinforcing steel, linear feet of pipe, hours of crane time, and the labor to install each one.

A detailed estimate starts with a **quantity takeoff**, the process of measuring every item of work from the drawings. How many cubic yards of concrete in the footings? How many square feet of formwork to contain it? How many pounds of rebar to reinforce it? Each quantity gets multiplied by a **unit cost** that includes material, labor, and equipment. The unit cost for placing concrete might be $180 per cubic yard, reflecting the cost of the concrete itself, the crew to place and vibrate it, the pump or crane to deliver it, and the finishing labor to strike it off. The sum of all line items gives the **direct cost** of the project.

Indirect costs cover everything that doesn't attach to a specific item of work but is necessary to run the job: the project manager's salary, the field office trailer, temporary power and water, insurance premiums, permit fees, dumpsters, and testing services. Indirect costs typically run 8 to 15 percent of direct costs, higher on complex urban projects and lower on straightforward rural work. The contractor adds a **markup** for profit, typically 3 to 10 percent depending on competition, risk, and the contractor's current workload. The total, direct costs plus indirect costs plus markup, becomes the bid price or the guaranteed maximum price, depending on the contract structure.

Estimating accuracy improves as design progresses and uncertainty shrinks. A conceptual estimate might carry a range of plus or minus 30 percent. A bid estimate aims for plus or minus 5 percent. The gap between these two numbers represents months of design work converting unknowns into specifications. Owners who set budgets based on conceptual estimates and then expect bid-level accuracy are a recurring source of conflict in the industry.

Scheduling converts the estimate's list of work items into a sequence plotted against time. The simplest format is a **bar chart** (Gantt chart), where each activity appears as a horizontal bar spanning its start and finish dates. Bar charts are easy to read but don't show the logical relationships between activities: they don't tell you that framing can't start until the foundation is complete, or that drywall can't begin until rough plumbing and electrical are inspected.

The **Critical Path Method** (CPM) addresses this limitation by modeling the project as a network of activities connected by logical dependencies. Each activity has a duration, and the dependencies define which activities must finish before others can start.

Finish-to-start is the most common relationship: Activity B can't start until Activity A finishes. **Start-to-start**, **finish-to-finish**, and **start-to-finish** relationships handle cases where activities overlap or must end simultaneously.

The CPM algorithm performs a **forward pass** through the network, computing the earliest possible start and finish for each activity, followed by a **backward pass** that computes the latest allowable start and finish without delaying the project.

The difference between an activity's earliest and latest start is its **float** (or slack): the amount of time the activity can slip without affecting the project completion date. Activities with zero float lie on the **critical path**, the longest continuous chain of activities through the network. Any delay to a critical-path activity delays the entire project by the same amount.

Identifying the critical path focuses management attention where it matters most. A 200-activity schedule might have 30 activities on the critical path and 170 with varying amounts of float. The project manager watches the critical path daily and applies resources, overtime, or resequencing to keep it on track. **Near-critical** activities, those with only a few days of float, also demand attention because small delays can shift them onto the critical path.

Resource loading assigns labor crews, equipment, and materials to each activity, revealing conflicts where the schedule demands more resources than are available. **Resource leveling** adjusts activity timing within available float to smooth the demand profile, avoiding the cost and disruption of hiring and firing crews week to week. A well-leveled schedule keeps crew sizes stable, which improves productivity because workers develop familiarity with their tasks and with each other.

Modern scheduling software like Primavera P6 or Microsoft Project handles networks with thousands of activities, multiple calendars (accounting for weather days, holidays, and shift patterns), resource constraints, cost loading, and earned value tracking.

Earned value management (EVM) compares the budgeted cost of work actually completed against the budgeted cost of work scheduled and the actual cost of work performed, producing metrics like the **schedule performance index** (SPI) and the **cost performance index** (CPI) that flag problems early enough to correct them.

Contract Types, Procurement, and Delivery Models

The contract between an owner and a contractor defines who bears which risks, how payment is computed, and what happens when things go wrong. Contract type selection is one of the most consequential decisions an owner makes, often more important than the choice of contractor, because it shapes incentives throughout the project.

A **lump sum** (or stipulated sum, or fixed price) contract pays the contractor a single agreed price for the defined scope of work. The contractor bears the risk of cost overruns: if materials cost more than estimated or productivity falls short, the contractor absorbs the loss. If costs come in below estimate, the contractor keeps the savings. Lump sum contracts give owners cost certainty and motivate contractors to work efficiently, but they also incentivize contractors to minimize scope, resist change orders, and bid aggressively on quantities they believe the designer underestimated.

A **cost-plus** contract reimburses the contractor for actual costs of labor, materials, equipment, and subcontracts, plus a fee for overhead and profit. The fee may be a fixed amount, a percentage of cost, or an incentive fee tied to performance targets.

Cost-plus contracts shift cost risk to the owner but give the owner flexibility to change scope without the adversarial change-order negotiations that characterize lump sum work. They work well on projects where scope can't be fully defined before construction starts, such as renovation of historic structures or emergency repairs.

A **Guaranteed Maximum Price** (GMP) contract is a hybrid: the contractor works on a cost-plus basis but guarantees that the total won't exceed a ceiling.

Savings below the GMP are typically shared between owner and contractor, creating an incentive for the contractor to control costs without the full adversarial dynamic of lump sum pricing. GMP contracts are common in **Construction Manager at Risk** (CMAR) delivery, where the construction manager is engaged during design to provide constructability input and then takes on the construction risk under a GMP once the design is sufficiently complete.

Unit price contracts pay the contractor a fixed price per unit of measured work: per cubic yard of excavation, per linear foot of pipe, per ton of asphalt. The total contract value adjusts as actual quantities vary from estimated quantities. Unit price contracts suit projects where the scope is well-defined in type but uncertain in quantity, such as highway and utility work where soil conditions and rock lines can't be known precisely until excavation begins.

Project delivery models define how the owner organizes the relationships among designer, constructor, and other parties. **Design-Bid-Build** (DBB) is the traditional model: the owner hires a designer to produce complete documents, then solicits bids from contractors and awards the contract to the lowest responsible bidder. DBB gives the owner complete control over design and competitive pricing on construction, but it separates design from construction knowledge, produces adversarial relationships when drawings contain errors or omissions, and delivers the project sequentially, with construction starting only after design is fully complete.

Design-Build (DB) assigns both design and construction to a single entity. The owner defines requirements in a performance specification or bridging document, and the design-builder produces the design and builds the project under one contract. DB reduces the adversarial gap between designer and constructor, enables fast-track delivery where construction begins before design is fully finished, and gives the owner a single point of responsibility. The tradeoff is reduced owner control over detailed design decisions and the need for the owner to define requirements clearly at the outset because changes after award are expensive.

Construction Manager at Risk (CMAR) brings a construction manager on board during design to advise on cost, schedule, constructability, and phasing. The CM then converts to a general contractor under a GMP when design reaches an agreed milestone. CMAR captures much of the collaboration benefit of design-build while preserving the owner's relationship with an independent designer.

Integrated Project Delivery (IPD) binds owner, designer, and constructor into a single agreement with shared risk and reward. Profits and losses are pooled, decisions are made collaboratively, and the traditional adversarial boundaries dissolve. IPD is the most collaborative model but also the least common, requiring a level of trust and cultural alignment that many organizations haven't developed.

Public projects in the United States have historically been required to use design-bid-build with award to the lowest responsible bidder, a procurement model that prioritizes cost competition and transparency. Legislation in most states now permits alternative delivery methods, and design-build has grown rapidly in public transportation, water, and facilities work.

Best-value procurement, which evaluates qualifications, technical approach, and price together rather than awarding solely on low bid, has gained acceptance as owners recognize that the lowest bid doesn't always produce the lowest total project cost.

Safety Management and OSHA Compliance on Jobsites

Construction is dangerous. In the United States, the construction industry consistently accounts for the most workplace fatalities of any sector, roughly 1,000 deaths per year, despite employing only about 5 percent of the workforce. The fatality rate per worker has declined dramatically over the past 50 years, thanks to regulation, training, technology, and cultural change, but construction remains a high-hazard occupation by any measure.

The **Occupational Safety and Health Administration** (OSHA) sets and enforces workplace safety standards. OSHA's construction standards appear in 29 CFR 1926, covering everything from excavation to scaffolding to fall protection to crane operation. Violations result in citations carrying penalties that range from thousands of dollars for serious violations to hundreds of thousands for willful or repeated offenses. Criminal prosecution is possible when willful violations cause a death.

The **Fatal Four** hazards account for more than half of all construction fatalities: **falls**, **struck-by** incidents, **electrocution**, and **caught-in/between** events. Falls from height kill roughly 350 construction workers each year in the United States.

OSHA requires fall protection at heights of 6 feet or more in general construction, with guardrails, safety nets, or personal fall arrest systems as the acceptable methods. Struck-by hazards include being hit by falling objects, swinging loads, vehicles, or rolling equipment. Electrocution occurs from contact with overhead power lines, faulty wiring, or energized equipment. Caught-in/between hazards include trench cave-ins, unguarded machinery, and collapsing structures.

Trenching and excavation receive special regulatory attention because cave-ins kill swiftly and without warning. OSHA requires protective systems for trenches 5 feet deep or more: sloping or benching the trench walls back to stable angles, shoring the walls with hydraulic or timber bracing, or shielding workers inside a trench box. A competent person must inspect the trench daily and after every rain event. Soil classification determines the required slope angles and shoring designs, tying geotechnical knowledge directly to life safety on site.

Scaffolding standards specify load capacities, platform widths, guardrail heights, access requirements, and inspection protocols. Scaffold erection and dismantling are themselves among the most dangerous activities on a jobsite, and OSHA requires a competent person to supervise all scaffold work. Aerial work platforms, commonly called boom lifts and scissor lifts, have largely replaced built-in-place scaffolding for many tasks, but they introduce their own hazards: tip-over on uneven ground, contact with overhead obstructions, and falls from the platform if the operator isn't wearing a harness and lanyard.

Crane safety involves engineering, inspection, and operational controls. Mobile cranes must be set up on firm, level ground with outriggers fully extended. Lift plans document the crane's rated capacity at the required radius and compare it to the weight of the load plus rigging. Capacity charts derate as radius increases, and exceeding the rated capacity is the primary cause of crane tip-overs. Tower cranes on high-rise projects require foundation design, climbing procedures, anti-collision systems when multiple cranes share airspace, and wind speed monitoring to determine when operations must shut down.

A project's **site-specific safety plan** translates OSHA's general standards into procedures tailored to the actual work. The plan identifies the hazards present on the project, the controls that will be applied, the training required for each trade, the emergency response procedures, and the responsibilities of each party. General contractors on large projects employ full-time safety managers who conduct daily inspections, run weekly toolbox talks (short safety meetings focused on the week's activities), investigate incidents, and track leading indicators like near-miss reports and safety observation counts.

Safety culture is the less tangible but equally important dimension. Projects where leadership treats safety as a core value rather than a regulatory burden consistently outperform projects where safety is an afterthought. Leading contractors measure safety performance not just by injury rates but by proactive metrics: the number of hazards identified and corrected before anyone is hurt, the percentage of workers who complete voluntary safety observations, and the speed with which reported concerns receive a response.

The economic case reinforces the moral one: every serious injury generates direct costs (medical bills, workers' compensation premiums) and indirect costs (lost productivity, schedule disruption, equipment damage, regulatory fines, litigation) that typically exceed the direct costs by a factor of two to four.

> **Does shifting risk through contracts actually reduce project uncertainty, or just redistribute its consequences?**

Chapter 10: Surveying and Geospatial Engineering

Nothing gets built in the right place without surveying. Every foundation, every road centerline, every bridge pier, every utility trench depends on measurements that tie the designed geometry to physical locations on the earth's surface. Surveying is the oldest branch of civil engineering, practiced in ancient Egypt to re-establish property boundaries after the Nile's annual floods, and it remains indispensable in an era of GPS satellites and laser scanners. The tools have changed beyond recognition, but the underlying problem hasn't: determine where things are, where they need to go, and whether they ended up where they were supposed to.

(Surveyors use a total station to precisely measure angles and distances, tying planned infrastructure layouts to exact physical coordinates on the ground so construction occurs in the correct location.)

Geospatial engineering extends surveying's reach by integrating measurement data with computational tools that store, analyze, and visualize spatial information at scales ranging from a single building site to an entire continent. Geographic information systems, remote sensing, photogrammetry, and satellite positioning have converged into a discipline that informs every phase of infrastructure development, from planning corridors through undeveloped land to monitoring the long-term deformation of completed structures. This chapter covers the measurement methods, the instruments, the positioning systems, and the data management frameworks that make all of it possible.

Plane and Geodetic Surveying Fundamentals

Surveying divides into two broad categories based on how it treats the shape of the earth. **Plane surveying** assumes the earth's surface is flat over the area being measured. For projects covering a few square miles or less, the curvature of the earth introduces errors smaller than the measurement precision, and the flat-earth assumption simplifies every calculation. Most construction projects, subdivision plats, topographic maps of individual sites, and boundary surveys operate in the plane surveying domain.

Geodetic surveying accounts for the earth's curvature and is necessary for projects spanning large areas: highway corridors hundreds of miles long, national mapping programs, satellite positioning networks, and any work that must tie into a regional or national coordinate system. The earth isn't a perfect sphere; it's an oblate spheroid, slightly flattened at the poles and bulging at the equator. Geodetic calculations use an **ellipsoid**, a mathematical surface defined by its semi-major and semi-minor axes, as a reference for latitude, longitude, and height. The most widely used ellipsoid today is the GRS 80 (Geodetic Reference System 1980), which underlies the World Geodetic System 1984 (WGS 84) used by GPS.

The **geoid** is a different reference surface: the equipotential surface of the earth's gravity field that corresponds roughly to mean sea level extended through the continents. The geoid is irregular, lumpy where mass concentrations pull it up and depressed where mass is deficient.

Orthometric height, the elevation used in engineering and shown on topographic maps, is measured above the geoid. **Ellipsoidal height**, measured by GPS, is measured above the ellipsoid. The difference between the two, called the **geoid undulation**, varies from place to place and must be applied whenever GPS-derived heights are used for engineering work. In the continental United States, geoid undulations range from about -8 to -53 meters, meaning the geoid sits below the ellipsoid by varying amounts. The National Geodetic Survey publishes geoid models (currently GEOID18) that provide the undulation at any point, allowing conversion between the two height systems.

Coordinate systems organize measurements into a usable framework. **Geographic coordinates** (latitude and longitude) locate points on the ellipsoid but aren't convenient for engineering calculations because a degree of latitude and a degree of longitude represent different distances, and those distances change with location. **Projected coordinate systems** map the curved ellipsoid onto a flat plane using mathematical projections that preserve certain geometric properties at the expense of others. The **State Plane Coordinate System** (SPCS) divides the United States into zones, each using a projection (transverse Mercator for narrow north-south zones, Lambert conformal conic for wide east-west zones) that limits distortion to less than 1 part in 10,000 within the zone. **Universal Transverse Mercator** (UTM) divides the globe into 60 six-degree-wide zones, each with its own transverse Mercator projection. Both systems give engineers Cartesian x-y coordinates in feet or meters that they can work with using ordinary plane geometry.

Horizontal control networks consist of monuments whose coordinates have been determined to high accuracy. These monuments serve as reference points from which all other measurements originate. The **National Spatial Reference System**

(NSRS), maintained by the National Geodetic Survey, provides the primary control framework in the United States. Vertical control is established through **benchmarks**, monuments whose elevations above the geoid have been determined by precise leveling. Control surveys connect project-specific monuments to the national framework, ensuring that the work can be related to neighboring projects, utilities, and property boundaries.

Measurement types in surveying reduce to angles and distances. Every position determination, whether by classical triangulation, modern total station traversing, or satellite positioning, ultimately derives from combinations of measured angles and measured distances. Classical surveys measured horizontal angles with theodolites and distances with steel tapes or stadia rods. Modern surveys measure angles electronically and distances with infrared or laser energy, but the geometric principles are identical.

Traverse is the fundamental survey procedure for establishing control. A traverse connects a series of points by measuring the angle and distance between consecutive stations. A **closed traverse** returns to the starting point or closes on a second known point, allowing the surveyor to compute a **misclosure** that reveals the accumulated measurement error. The misclosure is then distributed among the measurements by adjustment methods (compass rule, least squares) to produce the best estimate of each point's coordinates. Traverse accuracy is expressed as a ratio of the linear misclosure to the total traverse length: a 1:10,000 traverse means the error is 1 foot per 10,000 feet traversed, adequate for most construction work. High-order control surveys achieve 1:100,000 or better.

Leveling determines elevation differences between points. A **level** is an instrument that establishes a horizontal line of sight, and a **level rod** is a graduated staff held vertically on the point whose elevation is being determined. The surveyor reads the rod at a known benchmark, moves to the next point, reads the rod again, and computes the difference in elevation from the two readings. Successive setups extend the elevation determination over any distance. **Differential leveling** is the standard procedure; **trigonometric leveling**, which uses measured vertical angles and slope distances, is less precise but faster over rough terrain.

Total Stations, Levels, and Electronic Distance Measurement

The **total station** is the primary instrument of modern land and construction surveying. It combines an electronic theodolite for measuring horizontal and vertical angles with an **electronic distance measurement** (EDM) unit for measuring slope distances, all housed in a single instrument that records observations digitally and computes coordinates in real time. A surveyor sets the total station over a known point, sights a prism (reflector) held on the target point, and the instrument returns the horizontal angle, vertical angle, and slope distance in a fraction of a second. Onboard software converts these raw observations to horizontal distance, elevation difference, and three-dimensional coordinates.

EDM works by transmitting an infrared or laser signal to a retroflecting prism and measuring the time or phase shift of the return signal. Phase-comparison instruments modulate the carrier signal at known frequencies and measure the phase difference between transmitted and received signals, yielding distance with millimeter-level precision over ranges up to several kilometers. **Reflectorless** EDM transmits a laser pulse that reflects off any surface, not just a prism, allowing measurements to inaccessible points like building facades, bridge soffits, and cliff faces. Reflectorless range is shorter and accuracy is slightly reduced compared to prism-based measurement, but the ability to measure without a target makes it invaluable for as-built surveys and monitoring.

Modern total stations incorporate features that would have seemed extraordinary a generation ago. **Robotic total stations** include servo motors that lock onto a prism and track it as the rodperson moves, allowing a single operator to run the survey. The instrument follows the prism autonomously, measuring angles and distances continuously, freeing the surveyor to walk the site and collect data without returning to the instrument between shots. **Automatic target recognition** (ATR) uses an image sensor to find and center on the prism electronically, eliminating the need for the operator to manually point the telescope.

Levels remain essential for precise elevation work. **Automatic levels** (also called self-leveling levels) use a compensator mechanism that automatically corrects for small deviations from horizontal after the instrument is roughly leveled. They're rugged, simple to operate, and adequate for most construction leveling. **Digital levels** read a bar-coded rod electronically, eliminating the human reading error that limits optical instruments, and achieve precisions of 0.3 mm per kilometer of double-run leveling. **Laser levels**, both rotating and fixed-beam types, project a visible or detectable reference plane across a jobsite, allowing machine operators and laborers to grade, form, and place materials to a known elevation without a rodperson.

Data collection has moved entirely from field books to electronic data collectors and tablet-based field software. The surveyor codes each point with attributes (corner of building, edge of pavement, invert of manhole, top of curb) that feed directly into CAD and GIS software. Field-to-finish workflows automate the drafting of topographic maps: points coded as "edge of pavement" connect automatically into linework representing the road edge, manholes plot with the correct symbol, and contour lines generate from the collected spot elevations. This automation has compressed turnaround times from weeks to days for topographic surveys of moderate complexity.

Global Navigation Satellite Systems and RTK Positioning

Global Navigation Satellite Systems (GNSS) have transformed surveying by enabling three-dimensional positioning anywhere on earth at any time without requiring line of sight between stations. GPS (United States), GLONASS (Russia), Galileo (European Union), and BeiDou (China) together maintain constellations of over 100 satellites broadcasting signals that GNSS receivers use to determine position.

The fundamental principle is trilateration. Each satellite broadcasts a signal encoded with its orbital position (ephemeris) and precise transmission time. The receiver measures the travel time of the signal and multiplies by the speed of light to compute the range to the satellite. Ranges to four or more satellites, combined with the known satellite positions, yield the receiver's three-dimensional coordinates. In practice, the receiver also solves for its own clock error, which is why four satellites are the minimum rather than three.

Standalone GPS positioning using a single consumer-grade receiver achieves accuracy of 2 to 5 meters horizontally, adequate for navigation but useless for surveying. Accuracy limitations come from atmospheric delays (the ionosphere and troposphere slow and bend the signals), satellite orbit errors, multipath (signals reflecting off nearby surfaces before reaching the antenna), and receiver noise. Achieving survey-grade accuracy requires eliminating or modeling these errors, and two strategies dominate.

Differential GNSS (DGNSS) uses a reference station at a known location to compute corrections that are transmitted to the field receiver in real time. Because most errors are spatially correlated (the same atmospheric delay affects both the reference and the rover if they're close together), the corrections cancel most of the error, improving accuracy to sub-meter or decimeter levels depending on the technique. Simple code-based differential corrections are available from public and commercial services and are adequate for GIS data collection and resource-grade mapping.

Real-Time Kinematic (RTK) positioning achieves centimeter-level accuracy by tracking the carrier phase of the satellite signal rather than just the code. The carrier wavelength is about 19 centimeters for GPS L1, and by resolving the integer number of wavelengths between satellite and receiver (the **integer ambiguity**), the system measures range to a fraction of a wavelength. RTK requires a base station broadcasting corrections to the rover receiver, with baseline lengths typically limited to 10 to 20 kilometers before accuracy degrades. **Network RTK** extends the range by using a network of continuously operating reference stations (CORS) that model atmospheric errors across a region and deliver corrections valid anywhere within the network, eliminating the need for the surveyor to set up a project-specific base station.

RTK positioning has become the standard for construction surveying, boundary surveying, and topographic data collection where conditions allow. A surveyor with an RTK rover can occupy a point, collect a centimeter-accurate coordinate, and move to the next point in seconds, covering ground many times faster than a total station crew. **Machine control** systems mount GNSS receivers on bulldozers, graders, excavators, and pavers, feeding real-time position data to automated blade or bucket controls that shape the ground to the design surface without staking. Machine control has reduced the need for traditional grade staking on highway and earthwork projects and improved grading precision from tenths of a foot to hundredths.

GNSS has limitations. Signals require clear sky view, and tree canopy, buildings, deep excavations, and tunnels block or degrade reception. Urban canyons with tall buildings cause severe multipath. RTK initialization, the process of resolving the integer ambiguity, can be disrupted by obstructions and requires several seconds to minutes to reestablish. For these reasons, total stations and levels remain essential instruments, and most survey crews carry both GNSS and conventional equipment.

Static GNSS surveys achieve the highest positional accuracy, better than 1 centimeter horizontally and 2 centimeters vertically, by occupying points for extended periods (30 minutes to several hours) and post-processing the observations against reference station data. Static surveys are used for establishing control networks, monitoring structural deformation, and tying local project control to the national reference framework.

Photogrammetry, LiDAR, and Remote Sensing

Photogrammetry extracts three-dimensional measurements from photographs. Capturing overlapping images of the same terrain from different positions and measuring the parallax (the apparent shift in position of objects between images), photogrammetrists compute the three-dimensional coordinates of points on the ground. The method has been used since the early twentieth century for topographic mapping from aerial photographs and remains a primary means of producing large-area maps, orthophotos, and digital elevation models.

Traditional aerial photogrammetry uses calibrated cameras mounted in fixed-wing aircraft flying planned flight lines with 60 percent forward overlap and 30 percent sidelap between adjacent strips. Ground control points surveyed by GNSS or total station tie the photographic model to the coordinate system. Stereo pairs of photographs are processed in photogrammetric workstations that extract elevation data, trace contours, and digitize features.

Unmanned aerial systems (UAS, commonly called drones) have democratized photogrammetry by reducing the cost and complexity of aerial data collection. A small quadcopter carrying a consumer-grade camera can survey 100 acres in an hour, producing thousands of overlapping images that **Structure from Motion** (SfM) software assembles into dense three-dimensional point clouds, orthomosaic images, and digital surface models with centimeter-level resolution. Construction sites use drone surveys for progress monitoring, earthwork quantity computation, stockpile measurement, and as-built documentation at intervals ranging from daily to monthly.

LiDAR (Light Detection and Ranging) measures distance by emitting laser pulses and timing their return after reflecting off surfaces. Airborne LiDAR systems mounted in aircraft or drones emit hundreds of thousands of pulses per second, each returning a range measurement that, combined with the sensor's position and orientation (determined by GNSS and an inertial measurement unit), yields a three-dimensional point with centimeter-level accuracy. The resulting **point cloud** contains millions to billions of points describing the terrain and everything on it: ground surface, buildings, vegetation, power lines, bridges.

LiDAR's great advantage over photogrammetry is its ability to penetrate vegetation. A laser pulse hitting a tree canopy generates multiple returns: the first from the top of the canopy, intermediate returns from branches, and a final return from the ground surface beneath. By filtering for last returns, analysts extract a **bare-earth** digital elevation model even under dense forest cover, which photogrammetry cannot do. This capability makes LiDAR the preferred technology for floodplain mapping, forestry inventory, corridor surveys for highways and pipelines through wooded terrain, and archaeological survey in forested regions.

Terrestrial laser scanning (TLS) applies the same LiDAR principle from a ground-based tripod rather than an aircraft. A terrestrial scanner rotates a laser beam through 360 degrees horizontally and roughly 300 degrees vertically, collecting millions of points per scan and producing a detailed three-dimensional model of everything visible from the scanner position. Multiple scans from different positions are registered together to eliminate shadowed areas and build a complete model.

TLS is used for as-built documentation of complex facilities (refineries, power plants, historical buildings), structural monitoring, forensic analysis of accident scenes, and capturing existing conditions in renovation projects where original drawings don't exist or can't be trusted.

Remote sensing in the broader sense encompasses any technology that acquires information about the earth's surface without physical contact. Satellite imagery from platforms like Landsat, Sentinel, and commercial providers (Maxar, Planet) provides multispectral data covering visible, near-infrared, and thermal wavelengths. Civil engineers use remote sensing for land use classification, impervious surface mapping for stormwater analysis, monitoring construction progress on large projects, detecting ground subsidence using interferometric synthetic aperture radar (InSAR), and assessing damage after natural disasters.

Geographic Information Systems for Infrastructure

A **Geographic Information System** (GIS) stores, manages, analyzes, and displays spatial data. At its simplest, a GIS links a database of attributes (pipe material, diameter, installation date, condition rating) to geographic features (the line representing the pipe's location on a map). At its most powerful, a GIS performs spatial queries, network analysis, overlay analysis, and geostatistical modeling that reveal patterns and relationships invisible in tabular data alone.

Civil infrastructure is inherently spatial. Every pipe, road, bridge, building, parcel, utility line, and drainage feature has a location, and that location matters for design, construction, operation, and maintenance. GIS organizes this spatial information into **layers**, each containing a single category of features: one layer for water mains, another for sewer lines, another for parcels, another for contours, another for buildings. Layers can be combined, queried, and analyzed in ways that answer practical questions. Which water mains are within 500 feet of a proposed excavation? Which parcels fall within the 100-year floodplain? What is the shortest path through the road network from the fire station to a given address?

Vector data represents features as points, lines, and polygons with associated attribute tables. Fire hydrants are points. Roads are lines. Parcels and building footprints are polygons. **Raster data** represents continuous surfaces as grids of cells, each containing a value: elevation in a digital elevation model, land cover classification in a satellite image, temperature in a climate dataset. Both data types coexist in a GIS, and analyses frequently combine them.

Asset management is one of the most impactful GIS applications in civil engineering. Municipalities maintain thousands of miles of water mains, sewer lines, and roads, each with different ages, materials, condition ratings, and maintenance histories. A GIS-based asset management system tracks the location and attributes of every asset, schedules inspections and maintenance, prioritizes replacement based on condition and criticality, and projects capital needs over planning horizons of 20 to 50 years. Without spatial organization, this data is just a spreadsheet; with it, a utility director can see which neighborhoods have the oldest pipes, which mains have the most breaks per mile, and where the next capital investment should go.

Utility mapping and **subsurface utility engineering** (SUE) use GIS to document the location and depth of buried utilities: water, sewer, gas, electric, telecommunications, and storm drain. Accurate utility records prevent the excavation damage that causes gas explosions, water main breaks, and fiber optic outages. SUE quality levels range from D (existing records only, least reliable) through A (precise horizontal and vertical location determined by exposing the utility at test holes, most reliable). GIS stores all quality levels and flags areas where utility conflicts may affect proposed construction.

Transportation planning uses GIS for traffic analysis zone delineation, trip generation modeling, network assignment, and corridor evaluation. Environmental review uses GIS to overlay proposed alignments with wetland boundaries, endangered species habitat, cultural resource sites, and environmental justice communities. Stormwater management uses GIS to delineate watersheds, compute impervious cover percentages, model runoff, and identify locations for green infrastructure.

Web GIS platforms like Esri's ArcGIS Online and open-source alternatives like GeoServer make spatial data accessible to non-specialist users through web browsers and mobile apps. Field crews use mobile GIS apps to collect inspection data, update asset conditions, and report maintenance issues, with changes syncing to the central database in real time. Public-facing web maps let citizens view construction project impacts, planned road closures, and floodplain boundaries without visiting a government office.

Construction Staking and As-Built Documentation

Construction staking transfers the designed geometry from plans to the ground so that builders know exactly where to dig, form, pour, and place. It's the direct link between the engineer's design and the contractor's work, and errors in staking produce structures in the wrong location, at the wrong elevation, or with the wrong dimensions. Staking is typically performed by the project surveyor using a total station, GNSS rover, or both, and it continues throughout construction as successive elements require positioning.

Offset stakes mark points offset a fixed distance from the actual construction feature, because the feature itself (a footing excavation, a curb line, a pipe invert) will be disturbed during construction. A stake marked "10' RT" means the intended feature lies 10 feet to the right of the stake. **Cut/fill stakes** indicate how much earth must be removed or added at each point to reach the design elevation. A stake marked "C 2.3" means the existing ground is 2.3 feet above the design surface and must be cut. "F 1.8" means the ground is 1.8 feet below design and must be filled. These markings, written on wooden laths or flagged hubs, are the language field crews read to shape the earth.

Slope staking locates the points where cut or fill slopes intersect the existing ground surface, defining the limits of earthwork. The surveyor iterates between computing the theoretical slope intercept and measuring the actual ground elevation until the two converge. On highway projects, slope stakes march along both sides of the alignment at regular intervals, defining the earthwork prism that the grading contractor must move.

Line and grade staking for utilities establishes the horizontal alignment and vertical profile of pipes, conduits, and channels. The surveyor sets stakes or marks at manholes, bends, and intermediate points, with the cut to the pipe invert noted on each. The pipe crew uses a laser placed inside the pipe or a string line stretched between grade marks to guide installation, maintaining the designed slope to within fractions of an inch over hundreds of feet. Gravity sewers are especially sensitive to grade accuracy because they depend on slope to maintain self-cleansing velocity, and low spots in the profile trap solids and eventually block the pipe.

Building layout establishes the corners, column lines, and elevations of a structure. The surveyor sets control points beyond the building footprint, typically on **batter boards** (horizontal boards nailed to posts) that carry string lines defining the building lines and column center lines. Foundation contractors measure from these references to position formwork. As the structure rises, the surveyor transfers control vertically using optical plummets or total stations sighting vertical targets, ensuring that each floor aligns with the one below it and that the building remains plumb.

As-built documentation records what was actually constructed, which inevitably differs from what was designed. Changes occur throughout construction: utilities shift to avoid unforeseen rock, foundations deepen when the soil is softer than expected, road alignments adjust to accommodate a property owner, and structural members get substituted when the specified section isn't available. As-built surveys capture the final location and elevation of every significant feature, and as-built drawings (often called record drawings) update the design documents to reflect reality.

As-built data feeds directly into the GIS and asset management systems described earlier. A water main installed three feet from its designed location needs to be documented at its actual position, not its planned one, because the next crew that digs in the area will rely on the recorded location. The as-built survey is the last act of the construction surveyor and the first entry in the facility's permanent spatial record.

Building Information Modeling (BIM) extends as-built documentation into three-dimensional parametric models. A BIM model contains not just geometry but attributes: material properties, installation dates, manufacturer data, maintenance schedules. When combined with survey data from laser scanning or photogrammetry, BIM produces a digital twin of the completed structure that supports facility management, renovation planning, and structural monitoring for the life of the building. The integration of surveying, GIS, and BIM represents the current frontier of geospatial engineering, connecting measurement to design to construction to operation in a continuous digital thread.

In an era of highly precise digital measurement, what are the remaining sources of error that can still undermine a project?

Chapter 11: Urban Planning and Infrastructure Systems

Cities are infrastructure. Strip away the restaurants, the shops, the offices, and the apartments, and what remains is a skeleton of pipes, wires, roads, rails, and signals that makes urban life possible. Urban planning decides where things go and what rules govern them. Infrastructure engineering builds the systems that keep them running.

The two disciplines operate on different timescales, planning in decades and engineering in months, but they converge on the same question: how do you arrange and connect the physical components of a city so that millions of people can live, work, and move in reasonable comfort without the systems collapsing under the load?

The answer has never been static. Cities that thrived under one set of conditions have failed under another. The canal cities of the industrial revolution gave way to railroad cities, which gave way to automobile cities, each transition reshaping street widths, block sizes, building heights, and the infrastructure buried beneath. Today's cities face another transition as climate change, electrification, autonomous mobility, and sensor networks rewrite the assumptions that shaped the last century of urban development. This chapter covers the regulatory frameworks, the underground networks, the transit systems, and the emerging instrumentation that together define how modern cities function as engineered systems.

Zoning, Land Use, and Municipal Code Frameworks

Zoning is the legal mechanism by which a municipality controls what can be built where. A zoning ordinance divides the city's land area into districts, each with rules specifying permitted uses (residential, commercial, industrial, mixed), maximum building height, minimum lot size, maximum lot coverage, required setbacks from property lines, parking requirements, and density limits expressed as dwelling units per acre or floor area ratio (FAR). Zoning is the most popular tactic a city has for shaping its physical form, and its effects accumulate over decades as individual properties develop or redevelop in conformance with the rules.

The legal basis for zoning in the United States rests on the police power of state governments, delegated to municipalities through enabling legislation. The Supreme Court upheld zoning's constitutionality in the 1926 decision Village of Euclid v. Ambler Realty Co., establishing that separating incompatible land uses serves the public welfare. **Euclidean zoning**, named after that case, is the traditional model: rigid districts with single permitted uses. A residential zone permits only housing. A commercial zone permits retail and offices. An industrial zone permits manufacturing and warehousing. The boundaries between zones are sharp, and uses that don't fit the district require a **variance** (an individual exception granted by a zoning board) or a **rezoning** (a legislative change to the zoning map).

Euclidean zoning's rigidity has produced well-documented problems. Strict separation of uses forces people to drive between home, work, shopping, and recreation, generating traffic and automobile dependence. Single-family zoning districts that prohibit duplexes, apartments, and accessory dwelling units restrict housing supply and contribute to affordability crises. Minimum parking requirements consume land with asphalt, increase development costs, and spread buildings apart in ways that discourage walking. Maximum density limits in central locations push growth to the urban fringe, consuming agricultural land and natural habitat.

Reform zoning models address these problems in various ways. **Form-based codes** regulate the physical form of buildings, their height, setback, facade treatment, and relationship to the street, rather than their use. A building that fits the physical character of the district can house apartments above shops above a restaurant, all without a variance. **Mixed-use zoning** permits residential, commercial, and sometimes light industrial uses within the same district or the same building. **Transit-oriented development** (TOD) overlays increase permitted density within walking distance of transit stations, concentrating housing and employment where public transportation can serve them efficiently. **Inclusionary zoning** requires or incentivizes developers to include affordable housing units in market-rate projects.

Subdivision regulations govern how raw land is divided into buildable lots. They specify minimum lot dimensions, street widths and geometries, sidewalk requirements, utility easements, stormwater management, and dedication of land for parks and schools. Subdivision regulations work in concert with zoning: zoning says what can be built on a lot, and subdivision regulations determine the size and shape of the lots themselves and the infrastructure that serves them.

Building codes regulate the design and construction of individual structures. The **International Building Code** (IBC), adopted with local amendments by most U.S. jurisdictions, specifies structural requirements, fire protection, egress, accessibility, energy efficiency, and mechanical systems.

Related codes cover plumbing (IPC), mechanical systems (IMC), fire prevention (IFC), and energy conservation (IECC). Code compliance is verified through plan review before construction and inspections during and after construction. The code official, typically a municipal building department, issues a certificate of occupancy certifying that the completed building meets all applicable requirements.

Comprehensive plans (or general plans, or master plans) sit above zoning and codes as the policy framework that guides a community's long-term development. A comprehensive plan typically addresses land use, transportation, housing, economic development, parks and recreation, utilities, environmental protection, and community facilities over a 20-year horizon. Zoning ordinances and capital improvement programs are supposed to implement the comprehensive plan, though the relationship between plan and regulation varies widely in practice.

Capital improvement programs (CIPs) are the financial plans through which municipalities fund infrastructure construction and replacement. A CIP typically covers a five-year period and lists every planned project: road reconstruction, water main replacement, new parks, sewer extensions, bridge rehabilitation, and facility construction. Each project is tied to a funding source (general obligation bonds, revenue bonds, impact fees, state and federal grants, general fund revenues) and a schedule. The CIP is where planning meets engineering and finance, and it's the document that determines which infrastructure gets built and which gets deferred.

Integrated Utility Corridors and Underground Networks

Beneath every city street lies a tangle of infrastructure that most residents never see: water mains, sanitary sewers, storm drains, natural gas lines, electric cables, telecommunications conduits, steam pipes in some older cities, and increasingly, chilled water loops and fiber optic networks. These systems share the same limited underground space, and managing that space is one of the most challenging coordination problems in urban engineering.

Utility corridors are the volumes of ground assigned to different utilities within a street right-of-way. Standard practice places water mains on one side of the street, sanitary sewers and storm drains in the center or opposite side, gas on one side, and electric and telecommunications on another, each at specified depths and horizontal offsets.

Minimum separation distances between parallel utilities prevent accidental damage during excavation and avoid interference (water main leaks eroding the bedding around a gas line, for example). Vertical clearances at crossings follow hierarchy rules: generally, water above sewer, pressure pipes above gravity pipes, and newer installations yielding to existing ones.

The problem is that these idealized arrangements rarely survive contact with reality. A century of piecemeal installation, inadequate record-keeping, abandoned utilities left in place, and emergency repairs performed without survey documentation has produced subsurface conditions in older cities that no one fully understands. **Subsurface utility engineering** (SUE) addresses this through systematic investigation: reviewing existing records, using surface geophysical methods (ground-penetrating radar, electromagnetic pipe locators) to detect utilities without excavating, and ultimately exposing utilities at critical conflict points through vacuum excavation or test pits. The investment in SUE pays for itself by avoiding the construction delays, change orders, and utility strikes that result from incomplete subsurface information.

Utility tunnels (or utilidors) consolidate multiple utilities into a single accessible corridor. Instead of burying each utility independently beneath the street, they're installed on racks or in trays within a walkable or crawlable tunnel. Utility tunnels eliminate the need to excavate the street for maintenance or new installations, reduce conflicts between utilities, and protect the systems from frost, traffic loading, and accidental damage. They're common on university campuses, military bases, and in some international cities (Paris, Helsinki, Tokyo) but rare in U.S. municipal streets because of their high initial cost.

Trenchless technology reduces the disruption of utility installation by eliminating or minimizing open-cut excavation. **Horizontal directional drilling** (HDD) steers a drill head along a curved path beneath the surface, pulling pipe or conduit back through the bore after the pilot hole is complete. HDD routinely installs water mains, gas lines, fiber optic conduits, and electric cables beneath rivers, highways, railroad tracks, and sensitive surface features without disturbing them.

Microtunneling uses a remotely controlled tunnel boring machine to install pipes of 12 to 120 inches in diameter along precise alignments and grades, typically for gravity sewers where grade accuracy is critical. **Pipe bursting** replaces existing pipes by pulling a bursting head through the old pipe, fracturing it outward while simultaneously pulling the new pipe into the expanded bore. **Cured-in-place pipe** (CIPP) lining rehabilitates deteriorated sewers and water mains by inserting a resin-saturated liner that cures to form a new pipe within the existing one.

Stormwater and sanitary infrastructure in older cities often shares the same pipes through **combined sewer systems**, designed in the nineteenth century when the volume of both flows was manageable. Combined sewers carry dry-weather sanitary flow to the treatment plant, but during storms the combined flow exceeds plant capacity and overflows through designated **combined sewer overflow** (CSO) outfalls directly to receiving waters, discharging untreated sewage. Federal consent decrees have required cities with combined systems to invest billions in CSO control: building storage tunnels that hold overflow for later treatment, separating combined sewers into dedicated storm and sanitary systems, implementing green infrastructure to reduce stormwater entering the combined system, and expanding treatment plant wet-weather capacity.

Utility coordination during construction projects requires early identification of all affected utilities, notification of utility owners, relocation or protection of conflicting utilities, and sequencing of work so that utility relocations precede roadway or building construction. On major transportation projects, utility coordination can consume more schedule time and budget than the roadway construction itself. Dedicated utility coordinators manage the process, tracking each utility owner's design, permits, and construction through a master schedule that integrates with the prime contractor's work.

Public Transit Networks and Multimodal Connectivity

Moving large numbers of people through dense urban areas by private automobile alone requires more road and parking space than most cities can provide or afford. **Public transit** moves people in shared vehicles on fixed or semi-fixed routes, achieving passenger throughputs per lane that private cars can't match. A single lane of a freeway carries about 2,000 vehicles per hour. A single rail line carries 20,000 to 40,000 passengers per hour. The arithmetic dictates that cities above a certain density need transit to function.

Transit modes span a wide range of capacity, speed, cost, and operational characteristics. **Bus transit** operates on existing streets with minimal infrastructure investment: stops, shelters, and perhaps signal priority at intersections. Local bus service runs every 10 to 30 minutes at speeds of 10 to 15 mph, limited by frequent stops, traffic congestion, and boarding delays. **Bus rapid transit** (BRT) improves on local service through dedicated lanes, off-board fare collection, level boarding at stations, and signal priority, achieving speeds of 15 to 25 mph and capacities approaching light rail at a fraction of the capital cost.

Light rail transit (LRT) operates electric rail vehicles on dedicated rights-of-way that may run in street medians, in separate corridors, or in short tunnels through downtown cores. LRT carries 5,000 to 20,000 passengers per hour per direction at speeds of 20 to 35 mph, with station spacings of half a mile to a mile.

(This DART light rail train in downtown Dallas operates on a dedicated urban corridor.)

Heavy rail (metro, subway) operates high-capacity trains on fully grade-separated rights-of-way, typically underground in city centers and elevated or at grade in outlying areas. Metro systems carry the highest passenger volumes of any transit mode, 40,000 or more passengers per hour per direction, at speeds of 25 to 40 mph including station dwell times. **Commuter rail** serves longer-distance travel between suburbs and central cities, with station spacings of several miles and speeds up to 60 mph or more, often sharing tracks with freight railroads.

Transit network design determines routes, frequencies, and transfer points. Two competing philosophies shape network structure. **Radial networks** focus routes on the central business district, providing direct service from outlying areas to downtown but requiring transfers for trips between non-downtown destinations. **Grid networks** distribute routes evenly across the city, making transfers necessary for most trips but providing roughly equal access to all destinations. Most real networks combine elements of both, with radial trunk lines feeding a downtown core and crosstown routes providing lateral connectivity.

Frequency is the single most important determinant of transit usefulness. A route running every 30 minutes requires passengers to plan their lives around the schedule. A route running every 10 minutes or less is frequent enough that riders show up and go, treating transit more like a utility than a scheduled service. The threshold is often cited as 15 minutes or better during operating hours. Networks that concentrate service on fewer high-frequency routes generally attract more riders than networks that spread the same resources thinly across many low-frequency routes.

Multimodal connectivity links transit with walking, cycling, and private vehicles to create a complete trip chain. **Park-and-ride facilities** at suburban rail stations let commuters drive to the station and take the train downtown. **Bike-share systems** extend the walkshed around transit stations from a quarter mile (a comfortable five-minute walk) to a mile or more (a comfortable five-minute bike ride), dramatically increasing the population within easy access. **Pedestrian infrastructure**, including sidewalks, crosswalks, accessible curb ramps, pedestrian signals, and weather protection, determines whether the walk to and from transit is safe and comfortable enough to attract riders. The quality of the last-mile connection often matters more than the quality of the transit ride itself.

Transit-oriented development concentrates housing, jobs, and services within walking distance of high-quality transit stations, creating the density and mix of uses that support high ridership. TOD typically allows building heights and densities well above the surrounding zoning, reduces parking requirements (because residents and workers have transit access), and emphasizes pedestrian-friendly design at the street level. Successful TOD generates ridership that justifies transit investment and creates real estate value that helps finance it.

Fare collection has evolved from cash and tokens through magnetic-stripe cards to contactless smart cards and now to account-based systems where riders tap a bank card or phone and the system computes the optimal fare automatically. Open-payment systems reduce barriers to entry for occasional riders and eliminate the need to purchase and carry a transit-specific card. **Fare integration** across agencies (bus, rail, ferry) allows a single payment for a multi-modal trip, removing the financial penalty for transferring that discourages riders from using the most efficient route.

Smart City Sensors and Infrastructure Instrumentation

Civil infrastructure has traditionally been built, commissioned, and then left alone until something breaks or a scheduled inspection reveals deterioration. **Smart infrastructure** changes this by embedding sensors, communications networks, and data analytics into the built environment, enabling continuous monitoring, predictive maintenance, and real-time operational control.

Structural health monitoring (SHM) instruments bridges, buildings, dams, and tunnels with sensors that measure strain, acceleration, displacement, temperature, and corrosion in real time. Strain gauges bonded to steel members track stress levels under live load. Accelerometers measure vibration frequencies that shift when structural stiffness changes due to cracking or connection degradation. Tiltmeters on bridge piers detect foundation movement. Fiber optic sensors distributed along the length of a structural member measure strain continuously rather than at discrete points, revealing the location and magnitude of stress concentrations that point sensors would miss.

The value of SHM lies in what it enables: moving from time-based inspection (check every bridge every two years regardless of condition) to condition-based maintenance (focus attention on the structures showing signs of distress and leave healthy ones alone). A bridge that shows no change in its vibration signature over five years of monitoring is almost certainly in good condition and doesn't need the same inspection intensity as one whose frequencies have shifted.

The economic benefit compounds across a large inventory: a state transportation department maintaining 10,000 bridges can't afford to instrument all of them, but instrumenting the 500 most critical or most deteriorated ones focuses inspection resources where they matter most.

Water system instrumentation monitors pressure, flow, and water quality throughout the distribution network. Pressure sensors at key nodes detect pressure drops that indicate main breaks or excessive demand. Flow meters at district boundaries enable water balance calculations that reveal leakage rates. Acoustic sensors mounted on pipes listen for the characteristic sound signatures of leaks, locating them before they surface and cause damage. Water quality sensors measuring chlorine residual, turbidity, pH, and conductivity at points throughout the network detect contamination events or treatment upsets in near-real time, providing an early warning system that grab sampling at intervals of hours or days cannot match.

Traffic sensors count vehicles, measure speeds, and detect the presence of vehicles at intersections. **Inductive loop detectors** embedded in the pavement sense the change in inductance when a vehicle passes over them. **Video detection** uses cameras and image processing to count, classify, and track vehicles through intersections and along corridors. **Radar** and **LiDAR** sensors mounted on poles detect vehicles without being embedded in the pavement, simplifying installation and maintenance. **Bluetooth and Wi-Fi readers** detect the unique identifiers broadcast by drivers' phones and compute travel times between readers, providing segment-level speed data across the network.

Adaptive signal control uses real-time traffic data to adjust signal timing continuously, responding to actual demand rather than following fixed time-of-day plans. Systems like SCOOT and SCATS have operated for decades in cities worldwide, reducing delays by 10 to 25 percent compared to fixed-time plans. Newer systems incorporate machine learning to predict demand and pre-position signal plans for anticipated conditions.

Smart lighting replaces conventional streetlights with LED fixtures equipped with dimming controls, occupancy sensors, and communication radios. Lights dim to 30 or 40 percent of full output when no pedestrians or vehicles are detected and ramp up as activity approaches, cutting energy consumption by 50 to 70 percent while maintaining safety. The communication network built into the lighting infrastructure serves double duty as a backbone for other sensors: air quality monitors, noise sensors, gunshot detectors, and environmental stations that report temperature, humidity, and wind speed.

Digital twins integrate sensor data with three-dimensional models of infrastructure to create living representations that update in real time.

A digital twin of a water distribution system shows current pressures and flows overlaid on the pipe network, highlights anomalies, and runs simulations of what-if scenarios (a main break at a particular location, a fire demand at a specific hydrant) using current conditions as the starting point.

A digital twin of a bridge combines the BIM model with sensor data to visualize stress distributions, track cumulative fatigue damage, and project remaining service life. The twin doesn't replace engineering judgment, but it gives the engineer a richer, more current picture of the infrastructure's condition than periodic inspections alone can provide.

We have an example, simple digital twin on the following page.

Cybersecurity is the unavoidable companion of networked infrastructure. Every sensor, controller, and communication link that connects infrastructure to the internet also creates a potential entry point for malicious actors. Attacks on water treatment systems, traffic signal networks, and power grids have moved from theoretical to documented. Protecting smart infrastructure requires network segmentation, encrypted communications, access controls, intrusion detection, and a security-aware design philosophy that treats every connected device as a potential vulnerability. The engineering profession is still developing the standards and practices for securing infrastructure that was designed long before anyone imagined it would be connected to a global network.

Can cities ever be truly "optimized," or are infrastructure systems inherently constrained by competing human priorities?

This is a simulation of a cable-stayed bridge. Eight sliders control loading (live load, wind, temperature), geometry (span, tower height, cable tension), and material properties (stiffness, damping). The bridge SVG redraws in real time as parameters change. Six structural metrics update live (deflection, stress, safety factor, cable force, frequency, and thermal expansion) with an alert system that flags when any combination pushes into warning or critical territory.

The model simulates a cable-stayed bridge with 8 adjustable parameters across three categories:

Loading — live load, wind speed, and temperature drive the primary structural demands. Crank wind past 30 m/s and you'll see flutter warnings kick in; push load above 900 kN/m and the safety factor drops toward critical.

Structure — span length and tower height change the geometry of the visualization in real time, while cable tension directly controls how taut the stay cables appear and how much mid-span sag develops.

Material — deck stiffness and damping ratio affect how the structure responds dynamically. Low damping with high wind is where resonance risk builds up.

The computed outputs (deflection, stress, safety factor, cable force, natural frequency, and thermal expansion) update continuously as you adjust sliders, and the alert system flags when any parameter combination pushes into warning or critical territory based on typical Eurocode/AASHTO thresholds.

Chapter 12: Sustainability, Resilience, and the Future of the Built Environment

The civil engineering profession built the modern world, and in doing so it consumed extraordinary quantities of material and energy. Concrete production alone accounts for roughly 8 percent of global carbon dioxide emissions. Steel manufacturing adds another 7 percent. The construction and operation of buildings consume about 40 percent of all energy used worldwide. Roads, bridges, dams, water systems, and wastewater plants add to the total. For most of the profession's history, these costs were accepted as the price of progress, and the engineering challenge was limited to making structures safe, functional, and economical. That framing is no longer adequate.

Two forces are reshaping how civil engineers think about every project they undertake.

The first is **sustainability**: the recognition that the resources consumed and emissions generated by infrastructure cannot continue at current rates without degrading the natural systems that human civilization depends on.

The second is **resilience**: the recognition that the climate is already changing, that extreme weather events are intensifying, and that infrastructure designed for historical conditions may not survive the conditions ahead. These two imperatives don't always pull in the same direction, but they converge on the same conclusion: the next generation of infrastructure must be designed, built, and operated differently from the last.

This chapter covers the analytical frameworks, the rating systems, the design strategies, the digital tools, and the emerging technologies that define this transition. Some are already standard practice. Others are experimental. All of them will shape the profession for the next several decades.

Life Cycle Assessment and Embodied Carbon Accounting

Traditional engineering economics evaluates a project's cost at the moment of construction and perhaps its maintenance cost over a defined service life. **Life cycle assessment** (LCA) expands that evaluation to include every environmental impact associated with a structure from raw material extraction through manufacturing, transportation, construction, operation, maintenance, and eventual demolition or recycling. The result is a comprehensive environmental profile that reveals impacts invisible in a construction-cost-only analysis.

LCA follows the framework established by ISO 14040 and ISO 14044, which define four phases: goal and scope definition, inventory analysis, impact assessment, and interpretation. The **inventory analysis** catalogs every input (raw materials, energy, water) and output (emissions to air, water, and soil; solid waste) associated with each life cycle stage. The **impact assessment** converts the inventory into environmental impact categories: global warming potential, ozone depletion, acidification, eutrophication, smog formation, and resource depletion, among others. Each category aggregates many individual emissions into a single indicator. Global warming potential, for instance, converts methane, nitrous oxide, and other greenhouse gases into **carbon dioxide equivalents** (CO_2e) using factors that reflect each gas's radiative forcing over a chosen time horizon, typically 100 years.

For buildings and infrastructure, the life cycle divides into stages labeled A through D in the European standard EN 15978, which has become the global reference. **Stage A** covers production (A1-A3: raw material supply, transport, manufacturing) and construction (A4-A5: transport to site, construction processes). **Stage B** covers the use phase: energy for heating, cooling, and lighting (B6), water use (B7), maintenance, repair, replacement, and refurbishment (B1-B5). **Stage C** covers end of life: demolition, transport, waste processing, and disposal. **Stage D** accounts for benefits beyond the system boundary, such as the value of recycled materials or recovered energy.

Embodied carbon is the subset of LCA focused specifically on greenhouse gas emissions, and it has become the most widely tracked environmental metric in construction. Embodied carbon includes all emissions from producing and transporting materials, constructing the building, maintaining it, and eventually demolishing it, but excludes the operational energy emissions (which are tracked separately as **operational carbon**). The distinction matters because operational carbon can be reduced toward zero through efficiency and renewable energy, but embodied carbon is largely locked in at the moment of construction. Once the concrete is poured and the steel is erected, the emissions associated with producing those materials cannot be recovered.

The numbers are substantial. A typical mid-rise office building embodies 300 to 500 kilograms of CO_2e per square meter of floor area, with concrete and steel together accounting for 60 to 80 percent of the total. A highway bridge embodies thousands of tons of CO_2e. A water treatment plant embodies tens of thousands. As operational energy systems decarbonize through grid greening and building efficiency, embodied carbon's share of total life cycle emissions grows. For a high-performance new building connected to a clean electrical grid, embodied carbon can represent 50 to 80 percent of total life cycle emissions over a 60-year service life.

Whole-building life cycle assessment tools like Tally, One Click LCA, and the Athena Impact Estimator allow designers to model the embodied carbon of a proposed building during design, compare material options, and identify the decisions with the largest impact. Substituting fly ash for a portion of Portland cement in concrete, specifying recycled-content steel, using mass timber instead of concrete for floor systems, and minimizing material quantities through structural optimization all reduce embodied carbon measurably. The reductions from individual substitutions typically range from 10 to 40 percent per material, and the cumulative effect of multiple substitutions across the whole building can reach 30 to 50 percent compared to a conventional baseline.

Environmental Product Declarations (EPDs) are standardized documents that report the LCA results for specific products, analogous to nutrition labels for food. Concrete producers, steel mills, insulation manufacturers, and other building material suppliers publish EPDs that designers use to compare options and select lower-impact products. A structural engineer specifying concrete can compare EPDs from different ready-mix suppliers and choose the one with lower global warming potential, often without any change to the structural design.

Carbon budgets and **embodied carbon limits** are beginning to appear in building codes and procurement policies. The City of Vancouver requires whole-building LCA reporting for large projects. Several European countries set maximum embodied carbon intensities per square meter. California's Buy Clean Act requires EPDs for structural steel, concrete, and flat glass procured for state-funded projects and establishes maximum acceptable global warming potential values. These regulations are early and limited, but they signal a direction that will expand as jurisdictions gain experience with LCA data and as the industry's capacity to respond matures.

Green Building Rating Systems and Net-Zero Design

Green building rating systems provide structured frameworks for evaluating and certifying the environmental performance of buildings. They translate broad sustainability goals into specific, measurable credits that design teams pursue, third-party reviewers verify, and building owners publicize. The systems have driven market transformation by making sustainability visible, comparable, and marketable.

LEED (Leadership in Energy and Environmental Design), developed by the U.S. Green Building Council, is the most widely used rating system worldwide. LEED v4.1, the current version, organizes credits into categories: Integrative Process, Location and Transportation, Sustainable Sites, Water Efficiency, Energy and Atmosphere, Materials and Resources, Indoor Environmental Quality, Innovation, and Regional Priority. Projects earn points for achieving specific performance thresholds or implementing specific strategies, and the total determines the certification level: Certified (40-49 points), Silver (50-59), Gold (60-79), or Platinum (80+). Energy performance carries the most points, reflecting its outsized environmental impact, but water efficiency, materials, indoor air quality, and site sustainability all contribute.

LEED's influence extends beyond the buildings it certifies. The credit structure has shaped design practice industry-wide: strategies like daylighting, high-performance envelopes, efficient HVAC systems, low-flow plumbing fixtures, construction waste diversion, and commissioning were once exceptional and are now standard, in part because LEED credits rewarded them early and the market followed. Criticism of LEED centers on its complexity, its cost (certification fees and documentation effort), its emphasis on design-phase decisions rather than actual operational performance, and the gap sometimes observed between predicted and measured energy use in certified buildings.

Other rating systems address different markets, geographies, and priorities. **Green Globes**, administered by the Green Building Initiative, offers a streamlined alternative to LEED for commercial buildings. **BREEAM** (Building Research

Establishment Environmental Assessment Method), developed in the United Kingdom, is widely used in Europe and Asia.

Living Building Challenge, from the International Living Future Institute, sets the most demanding standard: net-zero energy, net-zero water, and strict material health requirements with no trade-offs or substitutions. A Living Building must generate as much energy as it consumes over a year, capture and treat all its own water, and exclude a list of toxic materials from its construction.

The **Envision** rating system, developed by the Institute for Sustainable Infrastructure, applies specifically to civil infrastructure rather than buildings. Envision evaluates highways, bridges, water treatment plants, airports, pipelines, and other infrastructure across categories including quality of life, leadership, resource allocation, natural world, and climate and resilience. It fills a gap that building-focused systems leave open, because the sustainability challenges of a highway interchange or a wastewater treatment plant differ substantially from those of an office building.

Net-zero energy design aims to produce a building that generates as much energy as it consumes annually, typically through a combination of aggressive energy efficiency (reducing demand to the lowest practical level) and on-site renewable energy generation (usually rooftop or parking-canopy photovoltaic panels) to meet the remaining load. The energy efficiency measures include high-performance building envelopes (continuous insulation, high-performance glazing, airtight construction), efficient HVAC systems (heat pumps, energy recovery ventilation, radiant heating and cooling), LED lighting with daylight harvesting controls, and plug load management. Energy use intensities below 25 kBtu per square foot per year are achievable for office buildings, compared to the national median of roughly 70 kBtu per square foot.

Net-zero carbon extends the concept to account for embodied carbon as well as operational carbon, requiring the building to offset or avoid more carbon than it emits over its full life cycle. This is a higher bar than net-zero energy because it forces attention to material selection, construction processes, and end-of-life considerations that energy-only targets miss. Few buildings achieve net-zero carbon today, but the framework is gaining traction as embodied carbon awareness grows.

Passive House (Passivhaus) is a performance-based design standard originating in Germany that targets extreme energy efficiency through five principles: superinsulated envelopes, high-performance windows, airtight construction, thermal-bridge-free detailing, and mechanical ventilation with heat recovery. Passive House buildings use 60 to 80 percent less heating and cooling energy than code-minimum buildings and maintain comfortable interior conditions with minimal mechanical equipment. The standard is rigorous and prescriptive about performance verification, requiring blower-door testing, energy modeling, and post-occupancy monitoring. Originally developed for the central European climate, Passive House has been adapted to hot, cold, humid, and mixed climates worldwide.

Climate Adaptation and Resilient Infrastructure Strategies

Sustainability addresses the causes of environmental degradation. Resilience addresses the consequences. A sustainable bridge minimizes the carbon emitted during its construction. A resilient bridge survives the flood that climate change makes more likely. Both matter, and both require engineering responses that differ from traditional practice.

Climate projections inform resilient design by estimating how temperature, precipitation, sea level, and extreme weather will change over the infrastructure's expected service life. A bridge designed today for a 75-year service life must perform through conditions projected for the year 2100. Global climate models, downscaled to regional and local levels, provide the temperature and precipitation scenarios that engineers use to update design assumptions. The challenge is that projections carry substantial uncertainty, especially at the local scale and for precipitation extremes, and engineers must decide how much conservatism to add when the science provides a range of possible futures rather than a single prediction.

Sea level rise threatens coastal infrastructure directly. Global mean sea level has risen roughly 8 to 9 inches since 1900 and is projected to rise another 1 to 4 feet by 2100 depending on emission trajectory, with higher values possible if ice sheet dynamics accelerate. Coastal roads, bridges, water treatment plants, wastewater outfalls, stormwater systems, and port facilities designed for historical tide levels face increasing frequency and depth of flooding. Adaptation strategies include raising infrastructure above projected flood levels, installing flood barriers and pump stations, relocating critical facilities landward, and in some cases accepting periodic inundation and designing for rapid recovery.

Extreme precipitation is increasing in intensity in many regions as a warmer atmosphere holds more moisture. A storm that was historically a 100-year event may become a 50-year or 25-year event within decades. Stormwater systems, culverts, bridges, and dams designed for historical rainfall statistics may be undersized for future conditions. Updating the design storm to reflect projected rainfall intensities is conceptually straightforward but practically difficult because infrastructure investments last for decades and updating every existing system is unaffordable.

Engineers increasingly apply **climate factors** that multiply historical rainfall depths by 1.1 to 1.3 (or more, depending on location and projection confidence) when designing new facilities, providing a margin against the intensification that climate science projects.

Extreme heat degrades infrastructure performance in ways that historical design didn't anticipate. Asphalt pavements soften and rut under sustained high temperatures. Railroad rails expand and buckle. Bridge expansion joints run out of travel. Power lines sag as conductors expand, reducing clearance and sometimes triggering wildfires. Concrete structures experience accelerated carbonation and reinforcement corrosion in hotter climates. Design responses include selecting higher-temperature-grade asphalt binders, increasing expansion joint capacity, specifying heat-reflective surface treatments, and incorporating higher temperature assumptions into structural thermal analyses.

Wildfire risk affects infrastructure in expanding wildland-urban interface zones. Utility poles, transmission lines, and communication towers in fire-prone areas require fire-resistant materials or burial. Water systems must maintain pressure during firefighting demand surges while their supply sources may be contaminated by fire debris and retardant chemicals. Evacuation routes must be designed with sufficient capacity and resilient enough construction to remain passable during a fire event.

Resilience frameworks organize these individual responses into systematic strategies. The National Institute of Standards and Technology (NIST) Community Resilience Planning Guide provides a six-step process for communities to assess their infrastructure's vulnerability to hazards, set performance goals, and develop implementation strategies. The framework recognizes that resilience isn't just about designing individual structures to withstand extreme events, but so that the community as a system, its transportation network, its water supply, its power grid, its communications, its emergency services, can continue to function or recover quickly after a disruption.

Redundancy, **robustness**, **resourcefulness**, and **rapidity** are the four properties that define resilient systems. Redundancy means multiple pathways exist so that the failure of one component doesn't disable the system. Robustness means individual components can withstand forces beyond their normal loading without catastrophic failure. Resourcefulness means the community has the materials, equipment, and organizational capacity to improvise solutions when systems fail. Rapidity means the time to restore function after a disruption is short enough that cascading consequences are contained. Designing for all four properties simultaneously is more expensive than designing for code-minimum performance, and the investment must be justified by the consequences of failure, which is why critical facilities (hospitals, emergency operations centers, water treatment plants) are designed to higher resilience standards than ordinary buildings.

Nature-based solutions use natural systems to provide infrastructure services and buffer against climate impacts. Coastal wetlands absorb wave energy and reduce storm surge flooding more effectively per dollar than many engineered sea walls. Urban forests reduce heat island temperatures, absorb stormwater, and improve air quality. Floodplain restoration reconnects rivers to their natural overflow areas, providing flood storage that protects downstream communities. Green roofs and bioretention systems reduce urban runoff volumes and peak flows. These solutions often provide co-benefits (habitat, recreation, carbon sequestration, aesthetic value) that engineered-only solutions don't, and they tend to be more adaptable to changing conditions because natural systems adjust dynamically rather than requiring human intervention to modify.

Building Information Modeling and Digital Twins

Building Information Modeling (BIM) is the process of creating and managing a digital representation of a facility's physical and functional characteristics. A BIM model isn't a drawing; it's a database shaped like a building. Every element in the model, every beam, column, wall, pipe, duct, and fixture, carries attributes beyond its geometry: material, weight, cost, manufacturer, installation date, fire rating, thermal conductivity, and any other property the project needs to track. Changes made to the model propagate automatically to every view, schedule, and quantity takeoff derived from it, eliminating the inconsistencies that plagued traditional drawing-based documentation.

BIM operates at defined **levels of development** (LOD) that describe the reliability and detail of model elements at each project stage. LOD 100 represents conceptual massing. LOD 200 adds approximate geometry and generic system types. LOD 300 includes precise geometry suitable for construction documents. LOD 350 adds connection details and coordination with adjacent systems. LOD 400 represents fabrication-ready geometry with complete detailing. LOD 500 is the as-built model, verified against field conditions. Not every element needs to reach LOD 500; the project's BIM execution plan specifies which elements reach which level at each milestone.

Clash detection is one of BIM's most immediate practical benefits. When architectural, structural, mechanical, electrical, and plumbing models are combined in a coordination model, software identifies locations where elements from different disciplines occupy the same space: a duct passing through a beam, a pipe intersecting a column, an electrical conduit running through a mechanical shaft that's already full. These clashes, which in traditional practice were discovered during construction and resolved through expensive field changes, are identified and resolved during design when changes cost a fraction of their construction-phase equivalent. A complex project might generate thousands of clashes in the initial coordination run and resolve them methodically over several weeks of design coordination meetings.

4D BIM adds the dimension of time by linking model elements to the construction schedule, producing animated simulations that show the building assembling itself in the planned construction sequence. Project teams use 4D simulations to identify spatial conflicts in the schedule (two cranes occupying the same airspace, a concrete pour blocked by incomplete formwork on an adjacent bay), to communicate the construction plan to stakeholders who can't read a CPM schedule, and to evaluate alternative construction sequences before committing to one.

5D BIM adds cost by linking model quantities to unit prices, producing cost estimates that update automatically as the design changes. When the structural engineer deepens a beam, the model recalculates the concrete volume, the formwork area, the reinforcing weight, and the associated costs without anyone manually revising a spreadsheet. The accuracy of 5D cost data depends on the quality of the unit prices and the completeness of the model, but even approximate 5D integration gives design teams faster feedback on the cost implications of their decisions than traditional estimating workflows allow.

Digital twins extend BIM from a design and construction tool to an operations and maintenance platform. A digital twin is a model connected to real-time data from sensors embedded in the physical asset, updating continuously to reflect current conditions rather than as-designed or as-built conditions. A digital twin of a building integrates BIM geometry with data from the building management system (HVAC operating parameters, energy consumption, occupancy counts), from structural health monitoring sensors (strain, vibration, displacement), and from maintenance management systems (work orders, inspection records, equipment condition).

The twin enables **predictive maintenance**: algorithms analyze sensor trends to forecast equipment failures before they occur, scheduling maintenance at the optimal time rather than on a fixed calendar or after a breakdown. An HVAC compressor showing gradually increasing vibration amplitude and discharge temperature is flagged for inspection weeks before it would fail catastrophically, avoiding the unplanned downtime and emergency repair costs that reactive maintenance produces.

For infrastructure beyond buildings, digital twins are being developed for water distribution networks (modeling pressure, flow, and water quality in real time), bridge networks (tracking structural condition across an inventory of thousands of bridges), rail systems (monitoring track geometry and predicting maintenance needs), and entire city districts (integrating traffic, energy, water, and building systems into a unified operational model). The computational and data management challenges are substantial, but the potential to optimize infrastructure performance, extend service life, and reduce life cycle costs is driving rapid development.

Interoperability remains the persistent challenge in digital infrastructure. BIM models are authored in proprietary software (Revit, Tekla, Bentley) that uses proprietary file formats. Exchanging data between platforms requires open standards like **Industry Foundation Classes** (IFC), an ISO-standard data schema for describing building and infrastructure elements. IFC is comprehensive but imperfect: some data loses fidelity in translation, and not all software supports all IFC entities equally. The **buildingSMART** alliance coordinates IFC development and promotes open BIM workflows, but the industry hasn't yet achieved the seamless interoperability that would allow any tool to read any model without data loss. Until it does, coordination workflows will continue to require agreed-upon platforms, careful data management, and human oversight at every exchange point.

Emerging Frontiers: 3D-Printed Construction, Modular Systems, and AI-Driven Design

The technologies in this section are at various stages of maturity, from proven-but-not-yet-mainstream to experimental-but-promising. What connects them is that each one challenges a fundamental assumption of traditional construction practice and offers the possibility of building better, faster, cheaper, or more sustainably than current methods allow.

3D-printed construction (additive manufacturing at building scale) deposits material layer by layer to form structural elements without formwork. Most construction 3D printing uses a cementitious material extruded through a nozzle mounted on a gantry or robotic arm. The nozzle traces the outline of walls and structural elements, building up height one layer at a time, typically at rates of 1 to 3 feet of vertical rise per hour. The result is a monolithic structure with curved and complex geometries that would be prohibitively expensive in conventional formwork.

Several companies have printed complete single-story houses in 24 to 48 hours of print time, with costs claimed at 30 to 50 percent below conventional construction for comparable floor plans. Multi-story printed structures have been demonstrated, including a two-story office building in Dubai and a three-story apartment building in Germany. The U.S. military has printed barracks and shelters in field conditions, demonstrating the technology's potential for rapid deployment in disaster relief and remote construction.

The engineering challenges are real and unresolved. Layer adhesion between successive passes affects tensile and shear strength across the print direction, creating anisotropic behavior that conventional concrete design doesn't account for. Reinforcement integration is difficult: conventional rebar can't be placed between printed layers the way it's placed in formed concrete, and alternatives (embedded cables, fiber reinforcement, post-tensioning, robotic rebar placement) are still in development. Material properties depend on print speed, nozzle geometry, ambient temperature, and time between layers, introducing variability that current testing standards don't address.

Building codes haven't caught up: no major code body has published prescriptive requirements for printed structures, and projects proceed under alternative means and methods provisions that require project-specific engineering justification.

Despite these limitations, 3D printing addresses genuine construction industry problems. Labor shortages in construction trades are chronic and worsening in many markets. Formwork represents 40 to 60 percent of the cost of concrete construction, and eliminating it changes the economics fundamentally. Geometric freedom allows optimized structural forms (curved walls, variable-thickness shells, topologically optimized shapes) that use less material than the rectangular prismatic forms dictated by flat formwork. The technology is unlikely to replace conventional construction wholesale, but it's likely to find durable niches in affordable housing, military and disaster response construction, and architecturally complex projects where formwork costs would otherwise be prohibitive.

Modular and prefabricated construction moves work from the job site to a controlled factory environment, where weather, lighting, tool access, quality control, and labor productivity are all superior to field conditions. **Volumetric modular construction** produces complete three-dimensional room-sized modules, fully finished with MEP systems, fixtures, and interior finishes, that are transported to the site and stacked or connected to form the complete building. **Panelized construction** produces flat wall, floor, and roof panels in the factory and assembles them on site, offering less factory completeness but easier transportation.

The productivity advantages are well documented. Factory construction achieves 2 to 3 times the labor productivity of site construction for comparable work, driven by repetition, stable working conditions, and the ability to work on multiple modules simultaneously. Schedule compression is dramatic: the factory produces modules while the site foundation is under construction, and the building stacks and connects in days or weeks rather than months. A 32-story modular hotel in New York City was assembled from over 900 modules in less than 90 days of on-site stacking, a schedule impossible with conventional methods.

Quality is generally higher in factory settings because work occurs on stable platforms at comfortable heights with consistent lighting, and because quality inspection happens at each station before the module moves to the next. Waste generation is lower because materials are ordered to exact quantities, cut on CNC equipment, and any scrap is sorted for recycling within the factory's waste stream. Weather delays are eliminated for factory work, and the proportion of work performed on site (where weather matters) shrinks from 100 percent to 20 or 30 percent.

The constraints are logistical and regulatory. Module size is limited by highway transportation regulations: maximum widths of 14 to 16 feet, maximum heights of 13.5 to 15 feet, and maximum lengths of 60 to 70 feet, depending on jurisdiction and route. These limits constrain room sizes, ceiling heights, and building layouts in ways that conventional construction doesn't face. Building codes have not fully embraced modular construction, and jurisdictions vary in their acceptance of factory inspection versus on-site inspection. The construction industry's fragmented structure, with local trades, local suppliers, and local relationships, resists the centralized factory model. And the upfront investment in factory facilities and tooling requires sustained volume to amortize, creating a chicken-and-egg problem for manufacturers entering new markets.

AI-driven design applies machine learning, generative algorithms, and optimization techniques to engineering design problems that have traditionally relied on human intuition and iterative manual analysis.

Generative design tools explore thousands or millions of design alternatives that satisfy specified constraints (load capacity, deflection limits, material budgets, spatial requirements) and present the designer with options optimized for chosen objectives (minimum weight, minimum cost, minimum carbon, maximum stiffness). The designer sets the rules; the algorithm finds solutions that a human working manually would never discover because the search space is too large to explore by hand.

Topology optimization is a specific generative technique that determines the optimal distribution of material within a defined design space. Given a volume, a set of loads, supports, and performance constraints, the algorithm removes material from locations where it contributes least to structural performance and retains it where it contributes most. The resulting forms often look organic, resembling bone structures or tree branches, because nature solves the same optimization problem through evolution. Topology-optimized structural nodes for steel connections have been 3D-printed in metal and installed in real buildings, demonstrating weight reductions of 50 to 70 percent compared to conventional welded plates while meeting the same load requirements.

Machine learning models trained on historical project data can predict construction costs, estimate project durations, identify risk factors for schedule delays, and flag design features associated with constructability problems. Natural language processing applied to specifications, RFIs, and change orders extracts patterns that help project teams anticipate issues before they escalate. Computer vision applied to jobsite photographs and drone imagery detects safety hazards, monitors construction progress against the BIM model, and identifies deviations from the design before they're covered by subsequent work.

Structural optimization using AI extends beyond topology to include sizing, shape, and layout optimization of entire structural systems. Algorithms that combine finite element analysis with evolutionary or gradient-based optimization can size every member in a steel frame to minimize total weight while satisfying all code requirements for strength, stability, and serviceability. The weight savings compared to a competent but non-optimized human design typically range from 10 to 20 percent, representing material and cost savings that multiply across a large building or a fleet of similar structures.

The professional and ethical implications of AI in engineering design deserve attention. Engineers bear legal and ethical responsibility for the safety of their designs, and delegating design decisions to algorithms raises questions about accountability, transparency, and professional judgment. A topology-optimized form that no human engineer would have conceived may perform superbly in the load cases it was optimized for but may lack robustness against unforeseen conditions that an experienced engineer would have anticipated intuitively. Current practice treats AI as a tool that augments human judgment rather than replacing it: the algorithm proposes, and the engineer evaluates, validates, and accepts responsibility. Professional licensing requirements, peer review practices, and codes of ethics all reinforce this division, and they're likely to evolve as the technology matures and the profession develops confidence in its capabilities and limitations.

The convergence of these technologies, 3D printing, modular construction, AI-driven design, digital twins, advanced materials, and sensor networks, points toward a future in which civil infrastructure is designed by algorithms exploring solution spaces no human could navigate, fabricated in factories or by robotic systems on site, monitored by sensors throughout its service life, and maintained proactively based on predictive analytics rather than fixed schedules or post-failure repairs. That future is arriving unevenly, faster in some sectors and geographies than others, and the transition will take decades rather than years.

But the direction is clear, and the engineers entering the profession today will spend their careers working at the boundary between what's established and what's emerging, applying judgment that no algorithm yet replicates to decisions that matter enormously for the safety, functionality, and sustainability of the world their work creates.

Is sustainable infrastructure primarily a technical challenge or a societal one, and what does that mean for the role of engineers?

Conclusion

Every structure you see around you, from the sidewalk beneath your feet to the skyscraper on the horizon, exists because someone figured out how to make it stand up, carry its loads, and serve the people who use it. That's what civil engineering is. It isn't one skill or one subject. It's a collection of disciplines that share a common purpose: shaping the physical environment so that communities can function safely, efficiently, and sustainably.

This book has walked through those disciplines one at a time, starting with the basics and building toward the complex systems that define modern infrastructure. The path began with statics, the language of forces and equilibrium that every other specialty depends on.

Without understanding how forces balance, how moments rotate, and how free-body diagrams reveal what's hidden, none of the later chapters would make sense. Statics is the grammar of the profession, and fluency in it separates engineers who understand their work from those who merely follow procedures.

From there, the focus moved to materials. Concrete, steel, timber, masonry, and composites each bring a distinct personality to the structures they form. Concrete is strong in compression and weak in tension. Steel is strong in both but vulnerable to corrosion and fire. Timber is renewable and surprisingly capable but sensitive to moisture and grain direction. Masonry is ancient and durable but brittle without reinforcement. Composites are light and corrosion-proof but expensive and unforgiving of design errors. Knowing these personalities is the foundation of every material selection decision an engineer makes.

Structural analysis and design took those materials and put them to work, showing how beams bend, how trusses distribute axial forces, how frames resist lateral loads, and how limit state design ensures that every member has an adequate margin of safety against every plausible combination of loads. The chapter on geotechnical engineering reminded us that every structure ultimately rests on soil, a material that no one manufactured and no one can fully control. Soil classification, effective stress, consolidation, shear strength, and foundation design connect the elegant calculations of structural engineering to the messy, variable reality of the ground.

Transportation engineering addressed the systems that move people and goods: pavements designed to survive millions of wheel passes, alignments shaped to keep vehicles safe on curves and hills, and traffic operations governed by the surprisingly predictable behavior of vehicles flowing through networks. Water resources engineering followed the path of water from rainfall through watersheds, channels, pipes, reservoirs, and stormwater systems, applying hydraulic principles at every scale from a rooftop gutter to a major river.

Environmental and sanitary engineering tackled the systems that keep urban populations healthy: drinking water treatment that removes pathogens and contaminants, wastewater treatment that returns used water to the environment without degrading it, solid waste management that contains the byproducts of modern life, and air quality controls that limit the emissions from the engines and factories that power the economy. Construction management bridged the gap between design and reality, covering the estimating, scheduling, contracting, and safety practices that determine whether a project gets built on time, on budget, and without harming the people who build it.

Surveying and geospatial engineering provided the measurement foundation that every other discipline relies on, from the total stations and GNSS receivers that locate points on the earth's surface to the GIS platforms that organize spatial data into systems capable of managing entire infrastructure networks. Urban planning and infrastructure systems zoomed out to the scale of the city, examining how zoning, utility corridors, transit networks, and sensor systems work together to make dense human settlement possible.

The final chapter looked forward. Sustainability and resilience aren't optional additions to civil engineering practice. They're becoming the framework within which all other decisions are made. Life cycle assessment, embodied carbon accounting, green building standards, climate adaptation, digital twins, 3D printing, modular construction, and artificial intelligence are reshaping how infrastructure is designed, built, operated, and maintained. Some of these changes are already routine. Others are still experimental. All of them will matter to anyone entering the profession in the coming decades.

If there's a single idea that ties all twelve chapters together, it's this: civil engineering is applied judgment. The equations, the codes, the software, and the standards provide structure, but they don't make decisions.

An engineer decides which soil model fits the site, which load combination governs the design, which material best serves the application, which alignment balances cost against safety, and which maintenance strategy keeps the system functioning over its intended life. Those decisions require more than technical competence. They require the kind of understanding that comes from knowing why the methods work, not just how to use them.

The built environment is the largest thing humanity has ever created. Its roads, bridges, buildings, dams, tunnels, water systems, and power networks represent trillions of dollars of investment and centuries of accumulated knowledge. Maintaining what exists, adapting it to changing conditions, and building what comes next will occupy civil engineers for as long as people live in communities and depend on infrastructure to support their daily lives. The principles in this book are the starting point for that work. They don't change with fashion or technology. Forces still balance. Water still flows downhill. Soil still consolidates under load. Materials still have strengths and weaknesses that must be respected. Understanding these fundamentals deeply and applying them wisely is what makes a civil engineer, and it's what makes the built environment safe, functional, and enduring for the people it serves.

End of Chapter Questions & Example Answers

Chapter 1: If civil engineers design systems that outlive them by decades, how should that time horizon change the way they make decisions today under uncertainty?

It forces conservatism, but not the paralyzing kind. Engineers designing for fifty or seventy-five year service lives can't predict future loads, climate conditions, or usage patterns with any precision. So they build in margins. They choose durable materials over cheaper ones. They design connections that can be inspected and replaced. They think hard about adaptability, asking whether a bridge deck can be widened, whether a treatment plant can accept a future module, whether a foundation can handle loads that don't exist yet. The time horizon also changes how engineers weigh cost. A decision that saves money today but creates a maintenance nightmare in thirty years is a bad trade, even if the spreadsheet doesn't capture it. The best practitioners treat uncertainty not as a problem to solve but as a permanent condition to design around.

Chapter 2: When a structure fails, how often is the root cause a misunderstanding of forces rather than a lack of strength, and what does that imply about engineering practice?

More often than most people assume. Forensic investigations repeatedly show that failures trace back not to weak materials but to load paths the designer didn't anticipate, connections that couldn't transfer forces the way the analysis assumed, or stability conditions that were overlooked entirely.

The Hyatt Regency walkway collapse in 1981 is a textbook case: the steel was strong enough, but a change in connection detail doubled the load on a critical hanger rod. The implication is that understanding how forces flow through a structure matters at least as much as sizing individual members. Engineering education rightly spends enormous time on free-body diagrams and equilibrium, not because the math is hard, but because the conceptual discipline of tracing every force to its reaction is what separates competent practice from dangerous guesswork.

Chapter 3: If every material has fundamental weaknesses, how does an engineer decide which trade-offs are acceptable in a given design?

Matching the material's strengths to the demands of the application and mitigating its weaknesses through detailing, redundancy, or maintenance. Concrete cracks under tension, so engineers reinforce it with steel. Steel corrodes, so they coat it, encase it, or specify corrosion-resistant alloys. Timber is vulnerable to moisture and biological attack, so they detail connections to shed water and specify preservative treatments. The decision isn't abstract. It depends on the loading, the environment, the design life, the maintenance budget, local material availability, and the skills of the local workforce.

Chapter 4: To what extent are safety factors a substitute for uncertainty, and where do they fall short?

Safety factors are blunt instruments. They work by applying a multiplier to loads or a reduction to capacity, creating a buffer between expected conditions and failure. This handles normal variability well. Where they fall short is in situations the engineer didn't model at all. A safety factor of 1.5 on wind load doesn't help if the real threat is a foundation failure nobody analyzed. Safety factors also can't compensate for construction errors, design omissions, or loads the code doesn't cover. They're necessary but insufficient, which is why good engineering also relies on redundancy, ductility, inspection, and professional judgment.

Chapter 5: Given that soil conditions are never perfectly known, how do engineers justify confidence in foundation designs?

They don't claim certainty. They claim sufficiency. A geotechnical investigation samples a tiny fraction of the ground, sometimes a few boreholes across thousands of square meters. Engineers interpolate between those points using geological judgment and conservative assumptions, then specify construction verification like proof-loading piles to confirm the ground behaves as expected. Confidence comes from layering enough independent checks that a surprise in one area gets caught by another.

Chapter 6: Should transportation systems prioritize efficiency or resilience, and how does that choice shape long-term urban development?

Both, but the balance matters enormously. Systems optimized purely for efficiency, maximum throughput at minimum cost, tend to be brittle. A single closure on a highway network with no redundancy paralyzes a region. Systems designed for resilience include alternative routes, modal options, and excess capacity that looks wasteful in normal times but proves essential during disruptions. The choice shapes cities over decades. Efficiency-first networks encourage sprawl along high-capacity corridors. Resilience-oriented networks, with grids, transit options, and connected local streets, produce more adaptable urban forms. Most real systems try to balance both, but the underlying priority is visible in the infrastructure long after the planners who chose it have retired.

Chapter 7: How should engineers design water systems when historical data is no longer a reliable predictor of future conditions?

This is one of the hardest questions in civil engineering right now. For over a century, hydrologic design relied on the assumption of stationarity, that the statistical properties of rainfall, runoff, and streamflow don't change over time. Climate change has broken that assumption. Engineers are responding in several ways: using climate-adjusted rainfall projections, designing for ranges rather than single values, building in adaptive capacity so systems can be expanded or modified, and shifting from purely structural solutions toward green infrastructure that absorbs variability. None of these fully resolves the problem. Designing water infrastructure today means accepting that the 100-year storm is no longer a fixed number and building systems flexible enough to handle the range of futures that current science considers plausible.

Chapter 8: Where should the line be drawn between engineering solutions and policy decisions in protecting environmental and public health?

Engineers can tell you what's technically achievable and what it costs. They can design a treatment plant to hit any contaminant target you set. But choosing that target, deciding how clean is clean enough, how much risk is acceptable, and who pays, is a policy decision that belongs to elected officials, regulators, and the public.

The line blurs in practice because engineers often draft the technical standards that become regulation, and because policy makers rely on engineering judgment to understand what's feasible. The healthiest dynamic is one where engineers provide honest technical options and consequences, and policy makers make the value judgments. Problems arise when either side tries to do the other's job.

Chapter 9: Does shifting risk through contracts actually reduce project uncertainty, or just redistribute its consequences?

Redistribute, mostly. A fixed-price contract doesn't make unforeseen soil conditions less likely. It just determines who pays when they appear. If the contractor bears the risk, they price it into their bid. If the owner retains it, they carry a contingency. The total uncertainty in the project stays roughly the same regardless of contract structure. What changes is behavior. Parties who bear risk invest in managing it, which can reduce the probability or impact of specific events. But badly allocated risk, assigning geotechnical risk to a contractor who has no control over subsurface conditions, for example, produces inflated bids, disputes, and adversarial relationships that often increase total project cost rather than reducing it.

Chapter 10: In an era of highly precise digital measurement, what are the remaining sources of error that can still undermine a project?

Plenty. Instrument precision is only one component of measurement accuracy. Errors still come from incorrect control point coordinates propagating through an entire survey network, from atmospheric conditions affecting GPS signals, from poor field procedures like setting up over the wrong point, and from human mistakes in data entry and file management. LiDAR and drone surveys produce millions of points, but if the ground control is wrong, every one of those points is wrong by the same amount. The most dangerous errors aren't the obvious ones. They're systematic biases that look plausible in the data and only reveal themselves when steel doesn't fit or a drainage system flows the wrong direction.

Chapter 11: Can cities ever be truly "optimized," or are infrastructure systems inherently constrained by competing human priorities?

TThey can't, and that's not a failure. Optimization requires a single objective function, and cities don't have one. Residents want affordable housing, short commutes, green space, quiet streets, vibrant commercial districts, and reliable utilities, and these goals routinely conflict with each other. A wider road moves more cars but makes the street less walkable. A denser zoning plan supports transit ridership but strains existing water and sewer capacity. Every infrastructure decision creates winners and losers, and the winners and losers change depending on which metric you use to measure success.

Infrastructure engineers work within this tension, not above it. A sewer upgrade competes with a park renovation for the same municipal budget. A transit line that serves one neighborhood bypasses another. Burying a utility corridor improves streetscape aesthetics but triples the cost and complicates future maintenance access. These aren't engineering failures. They're the reality of building shared systems in places where people hold genuinely different values and priorities.

The engineer's contribution isn't to optimize the city but to make the physical systems reliable, adaptable, and transparent about their trade-offs so that the political process can make informed choices. Good infrastructure doesn't resolve competing priorities. It gives communities honest options to choose between.

Chapter 12: Is sustainable infrastructure primarily a technical challenge or a societal one, and what does that mean for the role of engineers?

Both, but the societal dimension is bigger. Engineers already know how to build net-zero buildings, design low-carbon pavements, and treat water with renewable energy. The technical solutions exist or are emerging fast. What lags behind is the willingness to pay for them, the regulatory frameworks to require them, and the political consensus to prioritize long-term environmental performance over short-term cost.

Engineers who treat sustainability as purely a technical puzzle will keep producing elegant solutions that don't get built. The ones who make a difference understand that their role extends into advocacy, public communication, and helping decision-makers see that the upfront cost of sustainable design is almost always cheaper than the long-term cost of not doing it.

Terms and Definitions

- **Active Earth Pressure** - The minimum lateral pressure a soil mass exerts on a retaining structure when the structure yields slightly outward, allowing the soil to reach a failure condition. It governs the design of free-standing retaining walls.
- **Activated Sludge** - A biological wastewater treatment process in which microorganisms suspended in an aeration basin consume dissolved organic matter. The biomass is settled in a secondary clarifier and partially returned to the basin to maintain the microbial population.
- **Allowable Stress Design (ASD)** - A structural design philosophy that compares actual service-level stress in a member to an allowable stress derived from the material's yield or ultimate strength divided by a single safety factor.
- **Anisotropic** - Describing a material whose mechanical properties vary depending on the direction of loading. Wood is anisotropic because its strength and stiffness parallel to the grain differ substantially from those perpendicular to the grain.
- **Atterberg Limits** - The water content boundaries that define the behavioral transitions of a fine-grained soil. The liquid limit marks the transition from plastic to liquid behavior, and the plastic limit marks the transition from semi-solid to plastic behavior.
- **Bearing Capacity** - The maximum pressure a foundation can apply to the underlying soil without causing shear failure of the soil mass. It is computed from soil strength parameters, footing geometry, and overburden conditions.
- **Biochemical Oxygen Demand (BOD)** - The quantity of dissolved oxygen consumed by microorganisms while decomposing organic matter in a water or wastewater sample, expressed in milligrams per liter. It is the primary measure of organic pollution strength.
- **Bioretention Cell** - A shallow landscaped depression filled with engineered soil and vegetation that captures stormwater runoff, filters pollutants, and promotes infiltration. Also known as a rain garden.
- **Building Information Modeling (BIM)** - The process of creating and managing a digital representation of a facility that carries geometric, material, cost, and performance attributes for every element. Changes to the model propagate automatically to all derived views and schedules.
- **California Bearing Ratio (CBR)** - A laboratory test that measures a soil's resistance to penetration by a standard plunger, expressed as a percentage of the resistance offered by a reference crushed stone. It is used to evaluate subgrade strength for pavement design.
- **Capital Improvement Program (CIP)** - A multi-year financial plan through which a municipality schedules and funds infrastructure construction, replacement, and rehabilitation projects. Each project is tied to a funding source and an implementation timeline.
- **Cavitation** - The formation and violent collapse of vapor bubbles in a fluid when local pressure drops below the vapor pressure. In pumps, cavitation damages impeller surfaces and degrades performance.
- **Centroid** - The geometric center of a cross section, located at the point where the section would balance if made of a uniformly dense material.

Bending stress equals zero at the centroidal axis and reaches its maximum at the extreme fibers.

- **Clash Detection** - The automated process within a BIM coordination model that identifies locations where elements from different disciplines occupy the same physical space. Resolving clashes during design avoids costly field changes during construction.
- **Coefficient of Consolidation** - A soil property that governs the rate at which excess pore water pressure dissipates in a saturated fine-grained soil under load. It determines how quickly consolidation settlement occurs.
- **Cohesion** - The component of a soil's shear strength that is independent of the applied normal stress, arising from molecular forces between particles and from bound water. Clays exhibit cohesion; clean sands do not.
- **Combined Sewer Overflow (CSO)** - The discharge of untreated mixed stormwater and sanitary sewage from a combined sewer system directly to a receiving water body when flow exceeds the treatment plant's capacity during storm events.
- **Composite Beam** - A structural system in which a concrete slab is mechanically connected to a steel beam through shear studs, forcing both elements to deflect together and substantially increasing strength and stiffness compared to the bare steel beam alone.
- **Compression Index** - A soil parameter that quantifies the change in void ratio per log cycle of effective stress for normally consolidated clay. It governs the magnitude of primary consolidation settlement.
- **Consolidation** - The time-dependent process by which a saturated soil decreases in volume as excess pore water pressure dissipates and the applied load transfers from the pore water to the soil skeleton.
- **Critical Path** - The longest continuous chain of dependent activities through a project schedule network. Any delay to a critical-path activity delays the entire project by the same amount.
- **Darcy's Law** - The equation relating groundwater flow velocity to hydraulic conductivity and hydraulic gradient. It states that flow velocity equals the hydraulic conductivity multiplied by the gradient.
- **Dead Load** - The permanent self-weight of a structure and all fixed attachments, including the structural frame, cladding, flooring, roofing, and permanent mechanical equipment. It remains constant throughout the structure's life.
- **Design Speed** - The chosen vehicle speed that governs the minimum geometric standards for a highway, including curve radii, sight distances, and superelevation rates. All geometric criteria derive from this single parameter.
- **Digital Twin** - A continuously updated digital representation of a physical asset that integrates BIM geometry with real-time sensor data. It enables predictive maintenance, operational optimization, and scenario simulation using current conditions.
- **Disinfection Byproducts (DBPs)** - Chemical compounds formed when disinfectants such as chlorine react with natural organic matter in drinking water. Trihalomethanes and haloacetic acids are the two regulated classes.
- **Drilled Shaft** - A deep foundation element formed by drilling a cylindrical hole to the required depth and filling it with reinforced concrete. Also called a caisson or bored pile.
- **Ductility** - The ability of a material or structural member to sustain large inelastic deformations before rupture. Ductile failure provides visible warning, unlike brittle failure, which occurs suddenly.

- **Effective Stress** - The portion of total stress at a point within a soil mass that is carried by the grain-to-grain contact skeleton, computed as total stress minus pore water pressure. It controls both the strength and compressibility of the soil.
- **Embodied Carbon** - The total greenhouse gas emissions associated with producing, transporting, installing, maintaining, and demolishing a building or infrastructure element, excluding emissions from operational energy use.
- **Environmental Product Declaration (EPD)** - A standardized document reporting the life cycle assessment results for a specific construction product. EPDs allow designers to compare the environmental impact of competing materials.
- **Equivalent Single Axle Load (ESAL)** - A standard unit for quantifying pavement damage from traffic, representing the effect of one pass of an 18,000-pound single axle. Heavier axles cause exponentially more damage, following an approximate fourth-power relationship.
- **Euclidean Zoning** - The traditional zoning model that separates land into districts permitting only a single category of use, such as residential, commercial, or industrial. Named after the 1926 Supreme Court decision in Village of Euclid v. Ambler Realty.
- **Factor of Safety** - The ratio of a system's available resistance to the applied demand. In slope stability, it is the ratio of resisting forces or moments to driving forces or moments along a potential failure surface.
- **Fiber-Reinforced Polymer (FRP)** - A composite material consisting of strong fibers (carbon, glass, or aramid) embedded in a polymer matrix. FRP is used in civil engineering as reinforcing bars, structural wraps, and bonded plates.
- **Floor Area Ratio (FAR)** - A zoning metric defined as the total floor area of a building divided by the area of the lot on which it sits. It controls building bulk and density.
- **Flue Gas Desulfurization (FGD)** - An air pollution control system that removes sulfur dioxide from the exhaust of fossil-fuel combustion, most commonly using lime or limestone scrubbers. Removal efficiencies range from 90 to 98 percent.
- **Free-Body Diagram** - A sketch that isolates a single body or portion of a structure from its surroundings and shows every external force and reaction acting on it. It is the essential first step in any equilibrium analysis.
- **Froude Number** - A dimensionless ratio of flow velocity to the speed of gravity wave propagation in an open channel, equal to velocity divided by the square root of gravitational acceleration times hydraulic depth. Values below one indicate subcritical flow; values above one indicate supercritical flow.
- **Generative Design** - A computational design approach in which algorithms explore thousands of alternatives satisfying specified constraints and present solutions optimized for chosen objectives such as minimum weight, cost, or embodied carbon.
- **Geoid** - The equipotential surface of the earth's gravity field that corresponds approximately to mean sea level extended through the continents. Orthometric elevations used in engineering are measured above the geoid.
- **Geosynthetic** - A manufactured polymer product used in geotechnical and civil engineering applications. Types include geotextiles for separation and filtration, geogrids for reinforcement, and geomembranes for containment.

- **Guaranteed Maximum Price (GMP)** - A contract arrangement in which the contractor works on a cost-plus basis but guarantees that the total cost will not exceed a negotiated ceiling. Savings below the ceiling are typically shared between owner and contractor.
- **Hydraulic Jump** - A turbulent transition from supercritical to subcritical flow in an open channel, characterized by a sudden rise in water surface and substantial energy dissipation. Engineers use hydraulic jumps in stilling basins to slow high-velocity spillway flows.
- **Hydration** - The exothermic chemical reaction between Portland cement and water that produces calcium silicate hydrate, the binding compound responsible for concrete's compressive strength. Hydration continues for months after initial mixing.
- **Hydrograph** - A graph of discharge versus time at a specific point on a stream, showing the rise, peak, and recession of flow in response to a precipitation event.
- **Intensity-Duration-Frequency (IDF) Curve** - A graphical or tabular relationship showing the expected rainfall intensity for storms of various durations and statistical return periods at a specific location. IDF data are the starting point for most stormwater design calculations.
- **Leachate** - Contaminated liquid that percolates through solid waste in a landfill, dissolving or suspending organic and inorganic pollutants. Liner and collection systems prevent leachate from reaching groundwater.
- **Level of Service (LOS)** - A qualitative rating from A through F describing traffic operating conditions on a roadway or at an intersection. LOS A represents free-flow conditions; LOS F represents breakdown with demand exceeding capacity.
- **LiDAR** - A remote sensing technology that measures distance by emitting laser pulses and timing their return after reflecting off surfaces. Airborne and terrestrial LiDAR produce dense three-dimensional point clouds used for topographic mapping, as-built surveys, and structural documentation.
- **Life Cycle Assessment (LCA)** - A systematic methodology for evaluating the environmental impacts of a product or structure across all stages of its existence, from raw material extraction through manufacturing, use, and end-of-life disposal or recycling.
- **Limit State** - A condition beyond which a structure no longer satisfies its performance requirements. Strength limit states address collapse and failure; serviceability limit states address deflection, vibration, and cracking under normal use.
- **Liquefaction** - The sudden loss of shear strength in loose saturated sand or silty sand under cyclic loading, typically during an earthquake, caused by pore water pressure building to the point where effective stress approaches zero.
- **Live Load** - The transient, movable load associated with occupancy and use, including people, furniture, vehicles, and stored goods. Live loads vary over time and are reduced on large tributary areas to reflect the low probability of simultaneous full loading.
- **Load and Resistance Factor Design (LRFD)** - A structural design philosophy that applies separate load factors to each load type and a resistance factor to the nominal capacity, ensuring that factored demand does not exceed factored resistance.
- **Manning's Equation** - The empirical formula relating flow velocity in an open channel to the channel's hydraulic radius, slope, and roughness

coefficient. It is the standard method for computing uniform flow depth and discharge.

- **Mechanically Stabilized Earth (MSE)** - A retaining wall system that uses layers of tensile reinforcement embedded in compacted granular backfill behind a facing element. The reinforced soil mass behaves as a gravity structure.
- **Modulus of Elasticity** - The ratio of stress to strain in the linear elastic range of a material's behavior, representing its stiffness. Structural steel has a modulus of approximately 29,000 ksi; normal-weight concrete ranges from 3,000 to 5,000 ksi depending on strength.
- **Moment of Inertia** - A geometric property of a cross section that quantifies its resistance to bending about a given axis, equal to the integral of the square of the distance from the axis times the differential area. Units are length to the fourth power.
- **Net Positive Suction Head (NPSH)** - The absolute pressure available at a pump's suction inlet above the liquid's vapor pressure, expressed in feet or meters of head. Insufficient NPSH causes cavitation and pump damage.
- **Net-Zero Energy** - A building performance target in which the structure generates as much energy as it consumes over the course of a year, typically through a combination of aggressive efficiency measures and on-site renewable energy generation.
- **Orthometric Height** - Elevation measured above the geoid, representing the physical height above mean sea level. It is the elevation datum used in engineering design and shown on topographic maps.
- **Parallel Axis Theorem** - The principle stating that the moment of inertia of a shape about any axis equals its moment of inertia about the parallel centroidal axis plus the product of the area and the square of the distance between the two axes.
- **Passive House** - A performance-based building energy standard emphasizing superinsulated envelopes, airtight construction, high-performance windows, thermal-bridge-free detailing, and mechanical ventilation with heat recovery. Buildings meeting the standard use 60 to 80 percent less heating and cooling energy than code-minimum designs.
- **Plasticity Index (PI)** - The numerical difference between a soil's liquid limit and its plastic limit, expressing the range of water content over which the soil behaves plastically. High PI values indicate expansive, problematic clays.
- **Preconsolidation Pressure** - The maximum effective vertical stress a soil has experienced in its geological history. It marks the boundary between recompression behavior (below the value) and virgin compression behavior (above it).
- **Prestressed Concrete** - Concrete in which high-strength steel tendons are tensioned before or after the concrete hardens, introducing compressive stresses that offset the tensile stresses from service loads. Prestressing allows longer spans and thinner sections than conventional reinforcement.
- **Probable Maximum Flood (PMF)** - An estimate of the largest flood that could physically occur at a given site under the most severe meteorological conditions. It is used to size spillways for high-hazard dams.
- **Real-Time Kinematic (RTK)** - A GNSS positioning technique that tracks the carrier phase of satellite signals and applies corrections from a base station to achieve centimeter-level accuracy in the field. It is the standard method for construction surveying and machine control.

- **Resilient Modulus** - A measure of a soil's recoverable deformation under repeated cyclic loading, used to characterize subgrade support for pavement design. It better represents actual traffic loading conditions than the California Bearing Ratio.
- **Response Modification Factor (R)** - A coefficient in seismic design that reduces the elastic design force to account for a structural system's ductility and energy dissipation capacity. Higher R values reflect greater ductility and result in lower design forces but stricter detailing requirements.
- **Shear Strength** - The maximum shear stress a soil can resist along a failure surface, governed by the Mohr-Coulomb criterion as the sum of cohesion and the product of effective normal stress and the tangent of the friction angle.
- **Specific Energy** - In open channel flow, the total mechanical energy per unit weight of water measured relative to the channel bottom, equal to the flow depth plus the velocity head. It reaches a minimum at critical depth.
- **State Plane Coordinate System (SPCS)** - A set of projected coordinate systems that divide the United States into zones, each using a map projection that limits distortion to less than one part in 10,000. It provides Cartesian coordinates for engineering surveys.
- **Stopping Sight Distance** - The minimum distance ahead that a driver must be able to see an object in the roadway and bring the vehicle to a complete stop. It equals the distance traveled during perception-reaction time plus the braking distance.
- **Topology Optimization** - A computational technique that determines the optimal distribution of material within a defined design space by removing material where it contributes least to structural performance. The resulting forms often resemble organic structures.
- **Total Station** - An electronic surveying instrument that combines a theodolite for measuring horizontal and vertical angles with an electronic distance measurement unit, recording observations digitally and computing three-dimensional coordinates in real time.
- **Trenchless Technology** - Construction methods that install or rehabilitate underground utilities with minimal or no open-cut excavation. Techniques include horizontal directional drilling, microtunneling, pipe bursting, and cured-in-place pipe lining.
- **Undrained Shear Strength** - The shear strength of a saturated fine-grained soil loaded too quickly for excess pore water pressure to dissipate. It is independent of the applied normal stress and is used for short-term stability analyses.
- **Water Hammer** - A pressure surge that propagates through a pipe when flow velocity changes abruptly, as from a rapid valve closure or pump trip. The pressure rise can be severe enough to burst pipes and damage equipment.
- **Watershed** - The land area from which all precipitation drains to a single outlet point on a stream. Watershed characteristics including area, slope, soil type, and land cover determine the volume and timing of runoff.
- **Yield Stress** - The stress at which a material transitions from elastic to plastic behavior, sustaining permanent deformation. For structural steel, it is the primary strength parameter used in design.

Works Cited & Recommended Readings

Works Cited

American Concrete Institute. (2025). *ACI CODE-318-25: Building code for structural concrete; code requirements and commentary*. American Concrete Institute.

American Institute of Steel Construction. (2017). *Steel construction manual* (15th ed.). American Institute of Steel Construction.

American Society of Civil Engineers. (2022). *Minimum design loads and associated criteria for buildings and other structures* (ASCE/SEI 7-22). American Society of Civil Engineers.

American Society of Civil Engineers. (2024). *Flood resistant design and construction* (ASCE/SEI 24-24). American Society of Civil Engineers.

American Society for Testing and Materials. (2020). *Standard test method for compressive strength of cylindrical concrete specimens* (ASTM C39-20). ASTM International.

Association of American State Highway and Transportation Officials. (2018). *A policy on geometric design of highways and streets* (7th ed.). AASHTO.

Association of American State Highway and Transportation Officials. (2022). *AASHTO LRFD bridge design specifications* (10th ed.). AASHTO.

Chanson, H. (2004). *The hydraulics of open channel flow: An introduction* (2nd ed.). Butterworth-Heinemann.

Chow, V. T., Maidment, D. R., & Mays, L. W. (1988). *Applied hydrology*. McGraw-Hill.

Coduto, D. P., Yeung, M.-C., & Kitch, W. A. (2016). *Geotechnical engineering: Principles and practices* (3rd ed.). Pearson Education.

Das, B. M., & Sobhan, K. (2022). *Principles of geotechnical engineering* (10th ed.). Cengage Learning.

Davis, M. L., & Masten, S. J. (2024). *Principles of environmental engineering and science* (4th ed.). McGraw-Hill Education.

Ghilani, C. D. (2024). *Elementary surveying: An introduction to geomatics* (16th ed.). Pearson Education.

Halpin, D. W., Senior, B. A., & Lucko, G. (2018). *Construction management* (5th ed.). John Wiley & Sons.

Hibbeler, R. C. (2023). *Engineering mechanics: Statics* (15th ed.). Pearson Education.

Hibbeler, R. C. (2023). *Structural analysis* (11th ed.). Pearson Education.

International Code Council. (2024). *2024 International Building Code*. International Code Council.

International Organization for Standardization. (2006). *Environmental management, life cycle assessment: Principles and framework* (ISO 14040:2006). ISO.

International Organization for Standardization. (2006). *Environmental management, life cycle assessment: Requirements and guidelines* (ISO 14044:2006). ISO.

International Organization for Standardization. (2018). *Organization and digitization of information about buildings and civil engineering works, including building information modelling* (ISO 19650-1:2018). ISO.

Mannering, F. L., & Washburn, S. S. (2021). *Principles of highway engineering and traffic analysis* (7th ed.). John Wiley & Sons.

McCormac, J. C., & Brown, R. H. (2015). *Design of reinforced concrete* (10th ed.). John Wiley & Sons.

McCormac, J. C., & Csernak, S. F. (2023). *Structural steel design* (7th ed.). Pearson Education.

National Academies of Sciences, Engineering, and Medicine. (2023). *Highway capacity manual* (7th ed.). The National Academies Press.

Terzaghi, K., Peck, R. B., & Mesri, G. (1996). *Soil mechanics in engineering practice* (3rd ed.). John Wiley & Sons.

U.S. Green Building Council. (2024). *LEED v4.1 rating system*. U.S. Green Building Council.

Recommended Readings

Breyer, D. E., Fridley, K. J., Cobeen, K. E., & Pollock, D. G. (2014). *Design of wood structures, ASD/LRFD* (7th ed.). McGraw-Hill Education.

> The most thorough treatment of wood and timber design using both allowable stress and load-resistance factor methods, essential for engineers working with mass timber and conventional wood framing.

Ching, F. D. K., & Winkel, S. R. (2021). *Building codes illustrated: A guide to understanding the 2021 International Building Code* (7th ed.). John Wiley & Sons.

Stands out for its visual explanations of complex code provisions, making the IBC accessible to engineers, architects, and students who need to interpret requirements quickly. Particularly helpful for occupancy classifications, fire-resistance ratings, and egress requirements that are difficult to parse from code language alone.

Institute for Sustainable Infrastructure. (2018). *Envision: Sustainable infrastructure framework guidance manual* (3rd ed.). Institute for Sustainable Infrastructure.

Provides a structured sustainability rating system comparable to LEED but tailored specifically to highways, bridges, water systems, and other non-building civil assets.

Kavanaugh, B. F., & Mastin, T. B. (2014). *Surveying: Principles and applications* (9th ed.). Pearson Education.

A strong companion to Ghilani's text, offering additional coverage of construction surveying procedures, route surveying, and practical field techniques.

Nilson, A. H., Darwin, D., & Dolan, C. W. (2016). *Design of concrete structures* (15th ed.). McGraw-Hill Education.

Serves as a comprehensive alternative to McCormac for reinforced concrete design, with detailed treatment of prestressed concrete, two-way slabs, and seismic detailing.

Peurifoy, R. L., Schexnayder, C. J., Shapira, A., & Schmitt, R. L. (2018). *Construction planning, equipment, and methods* (9th ed.). McGraw-Hill Education.

Covers construction equipment selection, earthmoving productivity, and field operations in detail that complements the management focus of Halpin's text.

Transportation Research Board. (2016). *Highway safety manual* (1st ed., with supplement). American Association of State Highway and Transportation Officials.

Introduces predictive crash analysis methods that let engineers quantify the safety effects of geometric design decisions, an essential companion to the Green Book. Increasingly referenced in state DOT project development processes where safety performance justification is required for design exceptions.

Wight, J. K. (2016). *Reinforced concrete: Mechanics and design* (7th ed.). Pearson Education.

Takes a rigorous mechanics-based approach to reinforced concrete behavior, including advanced topics in shear friction, strut-and-tie modeling, and seismic design that go beyond introductory texts.

www.ingramcontent.com/pod-product-compliance
Lightning Source LLC
LaVergne TN
LVHW010624100826
845148LV00014B/3093